Time and Memory

A Primer on the Scientific Mysticism of Consciousness

Stephen Earle Robbins PhD

ISBN: 1468137492
ISBN13: 9781468137491

Stephen E. Robbins, Ph.D.
Golden Willows Farms
2750 Church Rd
Jackson, WI 53037

stephenerobbins.com

11/21/2011

Table of Contents

Introduction:
The Koans of Vision and Consciousness

The Quality that can be defined is not the absolute Quality.
The names that can be given it are not Absolute names.
Quality is the origin of heaven and earth.
It is all pervading.
Unceasing, continuous
It cannot be defined…

- *Zen and the Art of Motorcycle Maintenance,*
on the Tao Te Ching

If you will behold your own self and the outer world, and what is taking place therein, you will find that you, with regard to your external being, are that external world.
- Jakob Boehme (1575-1624)

Who is it that sees?
Zen Koan
- Bassui

Prologue

At age thirty-three, on a still, clear day, from a ridge overlooking the misty blue valleys of the Himalayas, Douglas E. Harding made the discovery of his life: he realized that he had no head. As he describes it:

> It was as if I had been born that instant, brand new, mindless, innocent of all memories. There existed only the Now, that present moment and what was clearly given in it. And what I found was khaki trouser legs terminating downwards in a pair of brown shoes, khaki sleeves terminating sideways in a pair of pink hands, and a khaki shirtfront terminating upwards in – absolutely nothing whatsoever! Certainly not in a head. (*On Having No Head*, p. 24)

And he continues:

> It took me no time at all to notice that this nothing, this hole where a head should have been, was no ordinary vacancy. On the contrary, it was very much occupied. It was a vast emptiness vastly filled, a nothing that found room for everything – room for grass, trees, shadowy distant hills, and far above them snow peaks like a row of angular clouds riding the blue sky. I had lost a head and gained a world. (*On Having No Head*, p. 24)

Bassui, a revered Zen master of old, had once given a koan for contemplation: Who is it that sees? A koan is simply a question, as, in another example, "What is your original face before you were born?" or "What is the sound of one hand clapping?" The "question" can be even only a syllable such as "Mu." Months or years of concentration upon this koan by a practitioner of Zen will, it is hoped, bring about the "answer," that is, the experience of enlightenment. Douglas Harding had experienced the answer to Bassui's koan. He had been enlightened. Simultaneously, he saw the truth of the words of the German mystic, Jakob Boehme (in the opening chapter quotes), for indeed, Douglas Harding *was* the external world. In that moment, the limited personality viewed as Douglas Harding had dropped away. It was, he said, "an end to dreaming." He had realized his true Self, his real Identity. He was Cosmic Being.

There is another world of practitioners, namely that of science and philosophy. In this world, the profound insight of a Harding, a Bassui or a Jakob Boehme in all likelihood would be received with a benign indifference. Richard Dawkins, the theorist of evolution, in his work, *The God Delusion*, sets the tone, seeing no need for a Designer of the universe, therefore no need for a God. He is heavily reinforced by philosopher Daniel Dennett (*Breaking the Spell* and *Darwin's Dangerous Idea*). Ray Kurzweil (*The Age of Spiritual Machines*) expresses the source of the indifference nearly directly. He sees no difference between ourselves and the robots we will (shortly) build based upon our ever growing understanding of the neural architecture of the brain, complete with any neural "God modules" for religious experiences which science will discover. "They [the robots] will believe they are conscious. They will believe they have spiritual experiences. They will be convinced these experiences are meaningful."[1] There are two lynchpins we can fix upon, then, that underlie this thinking in this particular community of thought. One is the theory of evolution. The other, which for Dawkins is an implicit support rather than explicit one, I will label as "the findings of neuroscience."

Evolutionary theory, as Dawkins has labored mightily to promulgate, is seen as stripping the Designer out of the universe. The great mechanisms of evolution – mutations and natural selection – provide a means to account for the development of every variety of form, of creature, of dinosaur, of microbe or giant squid that we see in the universe. There is no need to postulate a Designer or any form of consciousness for this vast variety of forms and creatures. And as Dawkins points out, it has been the supposed need of a Designer that has been the very strongest form of proof that there must be a God of this universe. If there is no God, no Universal Creative Mind, then we can only look upon the experience of a Harding with bemusement. The apparent content of his experience as the experience of identity with a Universal Mind or Universal Perception can only be a curious, hopefully harmless delusion. A class in Evolution 101 might be a useful corrective for those such as Harding when interpreting their supposedly mystical experiences. Or we might suggest a class in the "neuroscience of belief." In *Why God Won't Go Away: Brain Science and the Biology of Belief*, the authors make it very clear: "What we know beyond question is that the mind is essentially a machine designed to solve the riddles of existence, and as long as our brains are wired the way they are, God will not go away."[2] The neurologist, Kevin Nelson (*The Spiritual Doorway in the Brain*) details a theory where all spiritual experience – near death experiences, out-of-the-body experience, mystical experience – is simply a function of "activity in the primitive brainstem, working in tandem with the limbic system, the most ancient area of our recently evolved cerebral cortex."[3]

So the companion in arms to evolutionary theory is neuroscience. In the term "neuroscience," I include a conglomerate of sciences clustering around the problem of the functioning of the brain, particularly how the brain supports or "generates" consciousness. This includes artificial intelligence (AI), cognitive science, and the study of memory and perception. The most concrete form of the problem and nature of consciousness is vision or the process of visual perception. Simply and concretely, how we see.

The Problem of Vision

Let me be very concrete about the problem that psychology, neuroscience and even computer science have attempted to answer. Let us choose a simple event, perhaps a white coffee cup sitting on the surface of a table top with the brown coffee-liquid and cream being stirred by a spoon. This is the problem: how do we see this coffee cup as an object clearly external to our body, in three dimensions, with its white color and brown coffee and a spoon stirring in a gentle motion over time? Equivalently, we can ask, how do we explain the origin of the *external image* of the coffee cup?

Why is this a difficult problem? Because the picture science has gained of the neural processes of the brain seems to utterly contradict the essential experience – the experience of an image of the cup, clearly external to us, in space, with a brown and cream liquid surface being stirred. We can draw a simple picture of the dilemma. Consider Figure I.1. On the one side is the cup or for that matter any external object. The circle represents the brain. The arrows represent rays of light reflected from the cup which travel to the eyes and retinas. From this point, the light rays are transformed into neural impulses – chemical flows along the nerves – which travel to various processing areas in the brain. But in the brain, from the point in time that the light strikes the retinas, science discerns nothing but these chemical impulses traveling around. We see nothing that remotely resembles the image of the cup in the brain. Nor do we see any projections of light *backwards* to the cup. We see, as Jeff Hawkins (*On Intelligence*) describes it, only "a dark, quiet brain."[4]

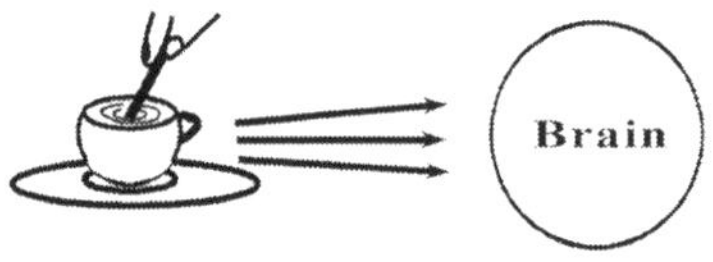

Figure I.1 Perception of a Coffee Cup

This is the great paradox of neuroscience. The best story our science can describe is a story of the details of neurons firing, of interconnections of neural groups to other neural groups, chemical reactions, electrical potentials, etc. There is a vast array of findings and knowledge on this neural structure, on the links and relationships of various neural groups or complexes in the brain. Nowhere in this extensive scientific story is there one clue as to how the image of the-cup-being-stirred is produced by the brain.

In the back of readers' minds likely lurks the question: "But how about computers? They generate images all the time. I see them on my laptop. Doesn't the brain do roughly the same thing?" Yes, the computer, via its software and electronics, can change the color values of the thousands of pixels or little dots that make up the projected screen image, creating scenes of oceans, or tigers, or movie stars. Or coffee cups. But the computer does not *see* these images. It does not see the coffee cup. It has merely executed processes or software that change the color values of thousands of separate pixels. Nothing more. This does not imply seeing. On the flip side, give the computer a camera and let it take a picture of the image of a coffee cup displayed on some other blind computer's screen. This recorded image, taken by the camera, must be transformed into digital values – patterns of 1's and 0's – represented by the states (say, "on" or "off") of the electronic components within the computer. The actual colors of the brown coffee, creamy cream and white coffee cup are gone. Only the representing bits or electronic components (some "on," some "off") remain. We thus have the same situation as the chemical flows of the "dark, quiet brain," only now we have the "dark, quiet machine." There is no clue here either how a computer could create and *see* the image of the external coffee cup.

This is such a critical point, let me approach it again, for the computer theorist says, "But look at all the computations taking place. These are computing the forms – the coffee cups and spoons – of the external world. And the results of these computations are what we are aware of." But, again, the "results of the computations" are only a coffee cup or a spoon from the viewpoint of an *external observer,* an observer *attributing* – as somehow existing across this mass of computations – the existence of a form or an image. In the computer there is only a massive series of bits flipping on and off. This occurs under the guidance of form-computing programs – themselves a series of bits or bit-based instructions being serially shoved, one after the other, into the registers of the computer. Where in all this is a global integrating awareness – a simultaneity enduring over time – that embraces this vast pattern of changing states and bits and says, "All this comprises a coffee cup and a stirring spoon?" We come back to this: there is no clue how the computer could actually see the coffee cup.

But the exponents of neuroscience are unfazed by this state of affairs. The problem goes today under the concept of the problem of "qualia." How does the external image of the world (whether visual or auditory) have its colors (redness, greens or blues) or its sounds (clanging, screeching, rustling)? In 1995, the philosopher, David Chalmers, famously declared this to be the all-important question: How, after you have described your computer architecture and your software programs, or your neural network, or your quantum computer, or whatever architecture, have you accounted for the qualia of the perceived world?[5] This was dubbed, "the hard problem." It has yet to be solved, despite volumes of discussion. Yet Nelson, in his *The Spiritual Doorway in the Brain,* also happily reducing all spiritual experience, to include Near Death Experiences (NDEs), to neural processes in the brain, spends perhaps three lines on the problem, dismissing it as something that eventually will be solved. The authors of *Why God Won't Go Away*, in their description of the neural architecture by which the brain supposedly generates our perceptions of the world, also effortlessly glide over the problem of the origin of the image of the external world; they also assume this is one minor problem that *will* go away. Hence they are quite certain of the truth of this statement: "*What we think of as reality is only a rendition of reality that is created by the brain.*"[6] All experience, let alone spiritual experience, "can be reduced to a fleeting rush of electrochemical blips and flashes, racing along the neural pathways of the brain."[7] The mystical experience of a Harding, we are assured, may indeed be that of a healthy mind "coherently reacting to perceptions that in neurobiological terms are absolutely real." Unfortunately, the two trillion hairy spiders which the schizophrenic sees crawling out of the walls are also "absolutely real in neurobiological terms." "Neurobiological reality" is a flimsy assurance to those seeking the meaning of mystical experience.

Practitioners of science, then, such as Dawkins, resolutely ignoring this minor koan of vision, are convinced that the current state of vast knowledge in neuroscience justifies a very significant conclusion: Since neuroscience identifies the entire process of perception as occurring within the brain, the conclusion is that it is the brain that must generate consciousness, it is the brain that generates the conscious visual perception, it is the brain that generates the visual image of the coffee cup with its colors, its form and the stirring spoon. Thus, since the brain is considered the sole and sufficient source of all these things, there is nothing else that science can say about the experience of Douglas Harding but this: it was simply an experience generated, strangely enough, by some chemical reactions and neural impulses, yet to be defined, by and within Douglas Harding's brain. The experience can have no significance beyond this. Combined with evolutionary theory, with the Designer-God or a Consciousness removed from the universe, this conclusion is reinforced.

This vision of the functioning of the brain has a side effect. It supports the extremely optimistic view that artificial intelligence (AI) and robotics will combine to create robotic intelligences equal to and greater than the humans that create them. Kurzweil is one of the very visible exponents of this, not to mention PBS specials interviewing MIT computer science professors. This is all based on the implicit view that we only need to replicate the hardware, the circuits and the processing architecture of the brain in silicon and in other artificial components to create intelligent devices with cognitive capacities beyond our own. This must be so, for the (imminent) final understanding and mapping of the brain's neural processing is seen as the entire key.

These two pillars of modern thought – neuroscience with its companion vision of robotic AI and evolutionary theory – deeply and symmetrically reinforce each other. Evolutionary theory sees the brain as naturally evolved via the process of natural selection without aid of a Designer. In turn the universe itself can be viewed, as physicist Seth Lloyd argues (*Programming the Universe*), as a vast (quantum) computing machine, seeded by random chance with a few simple programs which get the whole development going. This "universal computer" generates by simple computing processes in conjunction with natural selection all the forms, planets, galaxies, alligators and butterflies of the universe.

Quality – Lost

The origins of this view go deep. Outside the window of my little farm house where I now write, a coyote is trotting, passing by near the barn. I wonder what kind of journey he made to end up this far north. I am reminded of the narrative of another journey, a journey on a motorcycle, but not at all about a journey on a motorcycle. It is Robert Pirsig's classic, *Zen and the Art of Motorcycle Maintenance.* It is a book still deeply, profoundly relevant here. *Zen and the Art* was really a tale of one small fight in the battle for the nature of mind. The main character was not a coyote, but rather, Phaedrus – which is Greek for "wolf." "Phaedrus" was the name Pirsig gave to his own former personality, once a graduate student and teacher in philosophy, before Phaedrus became a casualty in the war, for Phaedrus suffered a nervous breakdown and eventually underwent electroshock therapy. *Zen and the Art* was his story.

Phaedrus had discovered a profound philosophy, an ancient philosophy, existing before Plato, before Socrates. Its fundamental vision was *Quality* or *Excellence.* The Greek heroes embodied it. Achilles, Hector, Ulysses were the embodiment of Quality – Excellence. The Tao, Phaedrus saw, was only another name for Quality. The very universe was Quality; the flow of time was intrinsically, Quality. Plato had abhorred this philosophy. The Sophists, the philosophers who espoused it, were transformed to a pejorative term (sophistry) implying deceptive logic. Plato subordinated Quality, which he called "the Good," to the "Truth," to "objective" reality, to reason, hence making Quality forever less than logic, less than reason, inferior to the intellectual mind. This was where the war was begun. Phaedrus, in a moment of stunning realization, saw the awful, profound consequences of Plato's victory. The seeds of the robotic vision of man were sown at that moment. This vision is incapable, intrinsically incapable, of incorporating Quality. But the essence of Man is Quality. The essence of the vast field of the universe of matter and its ever evolving, transforming flow in time, of which Man, like a vortex in a stream, is an integral part, is Quality. And by this, the essence of all our experience, all of our perception, all of our consciousness, is quality.

But in the reigning philosophical thought of our times, the framework that currently permeates all science, that permeates the Dawkins, the Dennetts, the Kurzweils, the universe has been stripped of quality. And in the computer model of mind, saturated with this Platonic framework, there seems no way to get it back. The philosophers, as I noted, coming only slightly to their senses about the computer model of man, now routinely term this fundamental, qualitative nature of our experience, of our vision and hearing, "qualia," and the computer model's inability to deal with this as "the problem of qualia" or the "hard problem." But being children of Plato, they are lost, helpless to contribute a solution. So the monstrosity that Phaedrus saw so intensely continues to stalk the world in far greater strength today. We see its image everywhere, in Arnold Schwarzenegger's Terminator, the R2D2s, Spielberg's machine-child in "AI," in the ever discussed threat of machines far surpassing, in the near future, the intelligence of man. It comes into the business world cloaked in the Capability Maturity Model, heavily sponsored by IBM, which specifies how humans should build computer software acting, for all practical purposes, as a group of robotic machines. It confuses education's approach to teaching, for it clouds the very nature of mind. It now permeates the discussion of spirituality – what there could be left of it – for spiritual experiences will eventually be generated by robotic machines that simply simulate, supposedly, what we humans in fact are. And the consequences of this deformed philosophy, championed by the winning side in a battle long ago, continue rolling along their dangerous course for the human race.

But...How do we see the coffee cup?

But there is that one nagging problem. Science has not one clue how we see the coffee cup on the table. Science does not know how Douglas

Harding saw the snow-peaks of the Himalayas. Science cannot explain conscious vision. In truth, science has set itself its own koan. Rather than ask, "Who is it that sees?" it asks, "How do we see?" or "What is the process by which we see?" But science has not been able to crack its koan. A Zen master would chastise this science – it has not meditated long and deeply enough. Science remains unenlightened. This unenlightened state is the Achilles heel of the practitioners of science and philosophy who feel they have a firm base for removing God from the universe, or for looking bemusedly at a Douglas Harding.

The fact is there *is* a scientific answer to science's koan. There *is* a solution for the philosophers. It was laid out in a model of perception developed by the French philosopher, Henri Bergson, in 1896. Subsequent developments in the theory and science of perception, in neuroscience and in physics have increasingly reinforced its plausibility. The pieces simply have not been put together within a single framework. This framework answers the question, "How do we see?" But simultaneously, it answers Bassui's koan, "Who is it that sees?" It turns out that we must answer this "who" question simultaneously in order to answer the question of "how." The one entails the other. The answer to this koan that I will give here will not be the experiential answer of a Harding, but it will, in its theoretical way, deeply support Harding's Zen-experience, and perhaps along the way impart the reader a glimpse of this experience.

I must prepare you for this: we will not, with the exception of a short discussion in the last chapter, be exploring various Zen experiences, mystical states, experiences in other realms of the universe. This will be a focus on the *everyday* mysticism of experience – "Channel Normal" if you will – and the reasons why it is so. Our everyday perception of the coffee cup is as mystical as a Zen enlightenment. It is not for nothing that the Buddha said, "Samsara *is* Nirvana." Samsara – the world of everyday objects, of coffee cups and spoons – is the mystical. This is simply not understood, either by non-mystics or by science. We will be seeing how mystical our Channel Normal really is, how far we are from robotic machines, and we will be getting glimpses of potentials we actually hold as beings, far beyond the limited and limiting conceptions of the machine model of mind. We will see why Channel Normal is not "generated by the brain," and in doing so, we will be laying the basis for an ability that might be possessed by that remarkable instrument that is the brain to perceive other realms as well. I could have left the spiritual relationship out of this book; I could have made it entirely and narrowly a treatment of a new model of the brain, of perception and memory. But the Dennetts, the

Dawkins, and so many others have dragged the spiritual nature of man into the fray, using their scientific findings on the brain and their neural models of cognition and memory as a bludgeon, that the spiritual question now screams to be addressed. Therefore I have chosen to declare the tie.

There have been many books advocating a larger view of mind – *The Holographic Universe*, *The Field*, *The Interconnected Universe*, the recent, *The Source Field Investigations*, and many more. But what is invariably lacking is a usable theoretical point of entry for the fundamental questions of consciousness, intelligence and memory being discussed by the academic world today. Simply saying that the brain must be a "readout" device for memories from the Zero-Point Field or that the brain "projects holographic images of the world" buys us little in the way of a usable entry point for current problems in memory or consciousness or cognition. In a word, I wish to give not only to the public, but to the academic world – the theorists of our everyday perception, consciousness and memory – the conceptual armament they need to move to a much less limited, far broader view of mind. We must clear the debris of misconceptions in cognitive science, consciousness theory, evolution and physics. The battle over the spiritual nature of man must first be won at the level of our understanding of the role of the brain and of the nature of our perception of Channel Normal – our everyday, ecological world.

End Notes and References

1. Kurzweil, Raymond (1999) *The Age of Spiritual Machines*, Viking, p. 153.
2. D'Aquili, E., Rause, V., Newton, A. (2001). *Why God Won't Go Away: Brain Science and the Biology of Belief.* New York: Ballentine, p. 35.
3. Nelson, Kevin (2011). *The Spiritual Doorway in the Brain: A Neurologist's Search for the God Experience*. New York, Dutton, p. 10.
4. Hawkins, Jeff. (2004). *On Intelligence*. New York: Times Books, p. 63.
5. Chalmers, D. (1995). Facing up to the problem of consciousness. *Journal of Consciousness Studies*, *2*, 200-219.
6. Hawkins, p. 35.
7. Hawkins, p. 143.

CHAPTER I

The Mystical Perception of Coffee Cups

The physical world is an infinity of movement, of Time-Existence. But simultaneously it is an infinity of Silence and Voidness. Each object is thus transparent. Everything has it own special inner character, its own karma of "life in time," but at the same time there is no place where there is emptiness, where one object does not flow into another.

- D. K., Canadian housewife, in *The Three Pillars of Zen*

Phaedrus felt that at the moment of pure Quality perception, or not even perception, at the moment of pure Quality, there is no subject and there is no object.

- *Zen and the Art of Motorcycle Maintenance*

The Koans of Life

It is my experience that one is prepared by Life. For what, one learns gradually, perhaps. All may be preparation. I was never much in high school mathematics. It just never penetrated, and my teachers had no clue how to get to me. I was in some other world. Of course the world I was in, in high school, was another world. My high school was a Franciscan seminary in Chicago. I apparently was always a somewhat mystical type. My Catholic imagination had been seized by Saint Francis when I was seven years old upon reading a book entitled, *The Perfect Joy of St. Francis*. A Franciscan – this is what I wanted to be. And my very religious parents faithfully shipped me down to Chicago from Minneapolis, with several other classmates of mine, via the Burlington Northern passenger train every school year. There I was immersed in English literature and creative writing, Latin, Greek, German, and French too. Not many people can say (or want to) that they have seven years of Latin. I read the New Testament in Greek, and Xenophon's tale of the journey of a Greek mercenary army through Asia, the *Anabasis* and the *Katabasis*. I dwelt in this environment through four years of high school and two years of college. I then entered the Novitiate, in theory, for what was supposed to be one year and a day, but ten months later, or two months and a day early, I said goodbye to the Franciscan world.

There is a book on the military academy at West Point entitled, "*The Long Gray Line*." I suspect there should also be a book entitled, "*The Long Brown Line*." My class was the last of the long brown line of Franciscan friars to experience the novitiate. What was now left of my sixty-plus high school class – twenty guys at age 20 or so – were cloistered in an enclosed monastic setting in the small farm town of Teutopolis, Illinois. The brick walls surrounding the enclosed space embraced enough room for a softball field, played with a sixteen inch ball, a small vineyard, a walkway of bricks comprising an oval circle about seventy five yards on the long axis with trees in the center, and plenty of southern Illinois heat. There, in the midst of summer, we donned the brown wool habit, the white cord with three knots, the long rosary of big brown beads, and sandals. There we endeavored to maintain a year long silence – except for breaks on Wednesday and Saturday afternoon when we played softball. We were roused out of bed in our tiny cells at 12am, slogged down the corridors to the little chapel and chanted Matins and Lauds in Latin for half an hour, then meditated for half an hour, then slogged our way to bed again at 1am. We were roused out again at 6am, plodded to the chapel and chanted Prime and Terce, attended mass, ate breakfast in the refectory in silence, did some work, studied, filed into the chapel at noon and chanted Sext and None, ate lunch in silence, did some work, went to chapel and chanted Vespers, ate dinner in silence, chanted Compline before making our way to our little cells on the second floor and bed at 9pm, woke up at 12am. Rinse and repeat. Did I mention we also walked the brick oval reciting the rosary on our big brown beads? Sundays were different. On Sunday we attended *two* masses. In our softball games, it was my great joy to pound the softball over the brick wall, which served as the border for left field, into the street. Right field was good too. Special permission was needed to retrieve the ball.

In the Franciscan world, it was common for friars to get advanced degrees. I had no doubt this is what I wanted to do. In this regime of chanting, prayer, reading philosophy, silence and softball, the desire came to me to do research and theory on "what it means to understand mathematics." My obtuseness in this subject always seemed curious. Something was wrong. My interest in psychology merged with my curiosity over the cause of my mathematical blindness. So ten months later, when I made my way back to Minnesota, it was psychology that became one of my majors. It was 1966. No longer with a draft deferment, I immediately found myself undergoing army medical exams in the draft induction process. I managed to gain one extra year by entering the University of Minnesota, one year in which to cram in the last two years worth of credits needed for a degree.

The psychology of 1966 was the realm of B.F. Skinner. It was dominated by rats – rats in their Skinner boxes, rats pushing levers, rats receiving food pellets, rats slaving under Skinner's "laws of reinforcement" – the roughly four laws that Skinner felt explained everything one needed to know about the mind, what there was of it in Skinnerian psychology. It was a bad time to be rat. It was a really boring time to study psychology. Only the last course I ever took, stuffed into my last summer session, given by a professor from the University of Michigan visiting the barren Skinnerian landscape of Minnesota, offered hope. Its subject was the great French theorist, Jean Piaget, and his theory of children's cognitive development. To me, it was the first glimmer that psychology could actually develop, or even care to develop, a theory of thought. It was actually dealing with what the brain might do. It was actually admitting there was a brain. This hope would keep me in the game. But there was something else. In describing the cognitive development of the infant, Piaget argued that initially the infant experiences no difference between himself and the world. For the child there is only the *subjective*. There are yet no "objects" distinct from the child as subject. Toys are part of the child. Moms are part of the child. Chairs are part of the child. There is only, in the whole world, one *subject* or self – the child. It is only over the course of cognitive development that this distinction – *subject* versus the *objects* of this world – is made.

I filed this away. It hurled me back to the philosophy course on "epistemology" I had received from Father "Fergie," my Franciscan philosophy professor, in my second year of college at the seminary, the last year of school before novitiate. Epistemology is supposed to be the study of "how we know what we know." How does a subject know about an object distinct from himself? How can one be certain of this knowledge? Fergie's fascination by this problem was impressive. It was a great course. I hadn't a clue what it was really all about. But it was filed away too like a koan – another one of those strange sayings of the Zen masters like, "What is the sound of one hand clapping?"

There would be no Zen Enlightenment however. It was 1967. The Vietnam War was raging. It was Uncle Sam who wished to give me his own experience of enlightenment. I would not return again to this subject for a few years. It would be very far from my mind.

Subject and Object

But it would all return. In the summer of 1970, I lived with three other Marine lieutenants. One was Ron, who was my officer basic school classmate, later a fellow officer in the third reconnaissance battalion in Vietnam and my roommate at the recon base in Quang Tri, a city just below the Demilitarized Zone. The other two were Duke and Jay. We shared two apartments on the top floor of a small yacht basin three feet from the inland waterway of North Carolina. A mile in the distance, the ocean roared. We were maybe half an hour from the Camp LeJeune Marine base. All of us had been infantry officers in Vietnam. All had been back for a little less than a year. All were slightly crazy.

The ritual, when we returned from base in the afternoon, was to run into the nearby town, usually on Duke's motorcycle. The mission was to find a wine with the highest alcoholic content possible. "Arriba" was our current favorite. We were then prepared for the night. Duke was the craziest. He had served as an infantry platoon commander for the standard thirteen months in Vietnam, then re-upped for an additional six months after the troops of his platoon, in appreciation for his leadership style, turned him on to marijuana. He now was a fan of LSD. Duke could be tripping, but you might never know.

Jay, who, with a Navy Cross, was the most highly decorated of the four of us, and myself, were a bit at arms length from any drug festivities. In my case, I had been in a profoundly involuted state ever since my return. I was a question to Life. I did not know why I had survived. I wondered profoundly – to what purpose? I was reading Teilhard de Chardin with his notion of the evolution of human consciousness, and the mystical Herman Hesse – *Demian, Siddartha, The Glass Bead Game*. A mutual girl-acquaintance of us all had given me a copy of Richard Bucke's *Cosmic Consciousness*. Poems were coming out of me that made little sense at the time. The Catholic Church and its doctrinal guidance, what there was of it, had been utterly abandoned *before* Vietnam. This latter came via an interesting move.

In Marine Officer Basic, in Quantico, across the hall from me and Hsieh Hua Cheng, my poet-officer Taiwanese-Marine roommate, was Mike O'Connor, a wrestler from the University of Iowa. One day, sitting in his room, towards the end of our six months of officer training, O'Connor, the Irishman essential, asked, "Are you going to go to confession before we go to Vietnam?" I said, "I don't know, why?" To which he said, "I think it would be just hypocritical. The only reason I would be going is because I'm scared. That just seems cowardly." "Yup," I agreed. Neither of us went to confession. I was pretty much done with the Church by that statement. They say there are no atheists in foxholes, but this scarcely does justice to the complexity of the souls of men in war. But O'Connor was killed on a recon patrol on the same day we both had sat on an LZ together talking, waiting for our respective helicopters to take our six-man teams into the middle of Vietnam's jungles, and the same day my own team got in a firefight, had a man killed and we were forced to make our escape.

There were enough of these experiences to put me in my "profound state of involution" in North Carolina. So very late one night, after our standard celebrations, I wandered out on the balcony of our second level dwelling, overlooking the sea, the sound of the waves in the distance, and the moon above, and found myself suddenly prone on the wooden deck, sobbing. With this began a mystical experience. I became aware of a Presence, a Being – a Being that had always been, a Being which "I" was, and which I had always been. This Being had done everything – good, bad, or everything under the Catholic definition of sin – that "I" had ever done. But there was no sin. Somewhere, I remember a statement of Thomas Merton to the effect that when one truly experiences contemplation, one realizes that there is no sin. There could be no sin, for this Being was the source of all action, and this Being was infinite. The ecstasy of the Love of this Being was so great, I could hardly bear it. The entire world was bathed in a literal glow of Love. I kissed the wooden lamp post, the wooden railings of the porch, the outside walls. They were glistening with a mystical glow. All was this Being, and this Being was Love. And all was Just. There was no injustice. What ever happened, to oneself, to another, it was absolutely just.

Duke came out and asked me if I was OK. I told him I was on a trip. This, he knew better. He looked quizzical, but he went back to bed. And in this profound state of ecstasy, I fell asleep, but only to dream. In the dream, I stood nearby and overlooking a long, long winding road, sensing this Being next to me. On the road, walking, filing by, was a vast stream of people. Suddenly, we – this Being and I – began racing swiftly up and down the length of the road, touching, traversing, melding with the minds of every individual moving in this vast stream. At this point I awoke.

Somewhere, in all this that night, I had seen that the relation between Subject and Object was not what psychology presumed it to be. I said to myself that if this relationship was redefined, an entirely new psychology of mind and cognition must result. Perhaps, I thought, vaguely, it would be profitable to pursue Piaget's conception of the initial oneness of subject and object in the child, studying the mechanism of the eventual partition of the child's undifferentiated world into "subject" and a world of separate "objects." This, I vowed my Being, was what I would dedicate myself to pursue when I could return to academics and graduate school.

It would have been a bit hard to function well in the Marine Corps very long in this state. Fortunately, Life intervened. News came within a week. We were being given an early out. Instead of February of 1971, I would be released from duty in September, 1970, just one more week. The day I got my release papers, I tried an exploration. Duke had long been advising that I should see the world on LSD. I had refused, feeling that if I had to report to the Camp LeJeune base the next day, any sense of time distortion would put me in an extremely paranoid state. Perhaps I would be late to work! Perhaps time would run backward! So that afternoon, in my new state of freedom, with Duke present in our yacht basin quarters, I gave it a try.

I quickly entered a state of fear; I was losing "myself," my identity. I told Duke this, and his response was, "What are you worried about losing?" This, strangely, did the trick. I entered the transformed perceptual world of LSD, with its vibrant colors and sounds, where wine glasses seemed extra translucent, where sounds seemed never ending with depth, where every thought quivered with layers of meanings that never ended. In this world, a little fly was buzzing around me. It was a very slow little fly, with wings flapping more slowly than the standard fly, and I waved slowly with my hand to move him away. I would remember this fly.

I was ready to return to Minnesota. All my belongings fit in my 1970 red Mustang Mach I, save for my little blue sailboat on the boat trailer attached. The Mach I would get there very fast. I would arrive just in time to begin the fall semester at the University of Minnesota.

The Koans of Bergson

It was fall of 1970. Poised just over a small hill of time, the computer model of mind was ready to sweep into psychology, birthing "Cognitive Science." The Skinnerian rats were doomed to unemployment. One of my first graduate courses that first semester was on the uses of mathematical group theory in psychology and the role it plays in Piaget's theory of cognitive development. The professor was Dr. William Bart, and he was quite fascinated with the notion of the mathematical group and its usefulness. We had to write a paper for the course, and I asked Dr. Bart if my topic could be, "The epistemology of Piaget." Now we have a pretty good idea where the "epistemology" thing came from. But Dr. Bart thought a moment and replied, "It would be better if you did a comparison of Bergson and Piaget." Sounded good to me. But who in heck was Bergson?

Henri Bergson, it turned out, was a French philosopher who wrote at the turn of the century. He was very famous circa World War I and his influence was very pervading. He was so famous that the French army in World War I appropriated one of his key concepts in the topic of evolution, an "élan vital" or vital force driving evolution. The French forces possessed the *élan vital.* Piaget was in fact a student (at least for a course or so) of Bergson.

It seemed Henri was always a bit special. He had shown his initial promise at nineteen, winning a prestigious national prize in mathematics. But he had turned instead to philosophy, causing his chagrined mentor, Desboves, to remark, "You might have been a mathematician, but you will be only a philosopher." Only a philosopher or not, he seems to have had a remarkable presence. The following description of Bergson was given by one of his classmates upon the later occasion of his reception into the French Academy:

> I recall quite clearly our first encounter. You were already famous then. You were always famous. I recall the fragile looking youth in those days with your slender, slightly

> swaying figure... You said little, but that little was uttered in a clear, sedate voice, full of deference to your companion's opinion, especially when you were proving him, in your quiet little way, and with that unconcerned air of yours, that his opinion was absurd. We had never seen a schoolboy so polite, and that made us regard you as somewhat different than ourselves – though not distant – you were never that and you have never been, but rather, somewhat detached and distinguished. From your personality emanated a singular, subtle charm; it was something quite subtle, even mysterious.[1]

He wrote and published his doctoral thesis in 1889, entitled, *Time and Free Will*. This was followed in five years, in 1896, with *Matter and Memory.* His most famous work, in 1907, was *Creative Evolution,* with its theory of the nature of evolution as being in fact the evolution of consciousness.[2]

I found a book, an exposition of Bergson's philosophy, by Wildon Carr. I dove into Carr's exposition. It was profoundly difficult to understand. I doubt with near certainty now that Carr truly understood the theory himself. He would not have been alone. But one thing arrested my attention immediately. It was this statement that Carr quoted from Matter *and Memory*: "*Questions of subject and object, in their distinction and their union, must be treated in terms of time, not space.*" [3] So there it was. Subject and object. But what did this mean – *treated in terms of time*?

Intuitively, I understood the warning on space. What I have said thus far about "subject and object" has been in terms of space. That coffee cup on my desk is located externally, in space, at a distinct spatial location from my body (the subject), and spatially distinct from the body. The musical sound is located externally in space, coming perhaps from the radio. It too is distinct from me. This is where the relation of subject and object merges with the problem of vision and perception.

A Return to the Great Koan of Vision

The coffee cup, the desk, the spoon, are of course, "objects." You perceive these objects. But how do you know they are really there? "You" are the knowing "subject." You are quite sure *you* exist. You feel the pain in your back perhaps if you have just gotten out of bed, perhaps even sleepiness, maybe your eyes are heavy. These internal sensations form a "you-ness" you are quite sure of. It is entirely "subjective," but very certain. But the cup and the book are *external* to you. How are you certain they exist? How do you know they are anything more than components of these internal sensations that comprise your subjective existence?

A certain philosophical position known as *idealism* would actually say that both book and cup *are* creations of the mind. All is idea. It is as though all is a dream. This position is not much in favor today. The opposite position is *realism*. Realism of course holds that the cup and book are very real and exist in an objective, physical, external world. However, there are two flavors of realism – direct and indirect. Direct realism holds unequivocally that you are looking at the actual cup and book in the objective external world. You have direct knowledge of their existence. Indirect realism is more cautious. Let's look again at our picture of the problem.

In Figure 1.1, the coffee cup is an external object, the circle is the brain – the home of the subject – and, as noted earlier, the arrows represent the light rays reflecting from the object to the brain. This is perception from a scientist's eye view. The rays continue through the retina, the light-energy is transduced and encoded within the neurons of the brain perceiving the cup into a neurally-based "representation" of the object – a representation consisting of chemical flows or electrons whirling that can look nothing like the cup. The processing of the physical light-energy ends in the visual sensory cortex – the appropriate processing area of the brain. As the philosopher, Mark Crooks notes, *there is no return of vision to the coffee cup.*[4] All perception then, even though of an *external* object, is occurring within the neural systems of the brain. This is the undeniable finding of neuroscience.

The paradox is clear: we cannot actually see into physical space or directly observe the cup, yet our experience is that of actually doing so. Yet the object appears located externally to the brain, in depth, in volume, in space. It is a disturbing, counterintuitive paradox. We are virtually wired to believe that perception is *direct,* that we truly see the coffee cup on the desk, right where it says it is. But Figure 1.1 stands in eloquent contradiction. In lieu of some incredible theory that replaces current science, we become philosophic ostriches, vainly kicking sand with a three-toed foot against the implications of Figure 1.1. The compromise, then, is *indirect realism*. The cup, the book, the Kleenex box on my desk are very real, but they are only *representations*, representations that look nothing like the box or the coffee

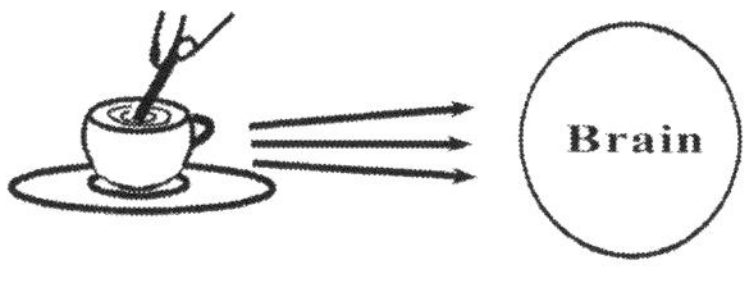

Figure 1.1 Perception of a Cup

cup, but that can be found somehow, and are supported somehow, in patterns of neural firing in the neural systems of the brain.

Patterns of neural firing in the brain. This is the rub. Whether the input is to the eye, or to the ear, or to the hand, it ultimately only becomes a pattern of neurons firing in cortical areas of the brain. The cortical pattern is spatial and temporal. It is spatial because the area of the cup is distributed simultaneously across an area of separate neurons in the visual cortex. It is temporal also. Whether the cup sits stationary or it moves, the patterns change over time – successively different neurons fire at different rates. If we are listening to a musical phrase, the firing of neural patterns in the auditory area of the cortex changes over time. There is yet a spatial aspect, for the inner ear or cochlea is laid out in space, as a line, and each frequency component of the sound causes a different (spatial) portion of inner ear or cochlea to vibrate. So for both sound and vision, we end with only spatial and temporal variation of patterns. As we saw Jeff Hawkins (*On Intelligence*) say, "there is no light in there," the "brain is dark inside," and quiet. Hawkins goes on to say:

> This is not to say that people or objects aren't really there. They are really there. But our certainty of the world's existence is based on the consistency of patterns and how we interpret them. There is no such thing as *direct perception*... Remember, the brain is in a dark quiet box with no knowledge of anything other than the time-flowing patterns of its input fibers.[5]

This is indirect realism. Note the denial of direct realism or direct perception. But how, from these firing patterns, do we get the white cup, with its brown coffee and rising steam, located *externally,* sitting on the desk top with spoon nearby, all spreading out before us? We may be convinced that the cup is real. The neuronal firing pattern for the cup may be indeed very consistent, occurring each time the cup is there. But how does the brain know? How do I, the subject, know? Yes, I can reach out and touch it, but what if I could not? And even if I touch it, how is this touch sense any different? It is just another brain-generated "image," a touch image or kinesthetic image. All is only firing patterns in the dark, quiet brain. Curiously, then, indirect realism offers only a *faith* in realism – the actual objective existence of the external world. It is a hair's breadth – a microscopic, in practicality, invisible hair – from saying this: all is only idea.

Current robotics and AI vision systems are being programmed to process information received from their external environment. The robot is given a camera which picks up the light in the surrounding world. The information in the light, the pattern, is again transduced or encoded, this time to on/off electric switches representing "bits," where each bit is either a 0 or 1. Just as the brain transduces the light to neural firing patterns, the robot takes the light patterns and encodes them to patterns of bits, 000110001, 10010011, etc., flipping the bit states from 0 to 1 – changing bit patterns in the "dark quiet world" of the robotic computer-brain. The robot can compute action from the information encoded in these bit patterns. The robot can *act* as though it sees. It can make its way along corridors or even, with new improvements, a little rougher terrain. But it is little different than a congenitally blind man receiving instructions: go left, go forward two steps, go one step right. There is action, but there is no vision, no image of the external world, only the changing patterns of bits. Though the movies depict Arnold, the Terminator, with a display screen of the external world complete with little digital readout numbers in red flicking across the bottom, there is no theory as to how Arnold-the-Terminator could see this readout screen and its image. Arnold, the Terminator, is blind as a bat. But he acts convincingly. He can defeat other robots. He can spit out one liners. A TV set, guided by its digital signals, paints light patterns on its screen. The TV does not see the screen, it does not see the image, we do, and it is *our* ability to see the image, how we do so, that is the problem. Arnold's picture display is just a mini-TV set, a part of Arnold, and Arnold too has no ability to see the image digitally painted on his screen. AI and robotics have not one thing to contribute to the solution on how the brain creates an image of the external world, or where the image comes from. Both are dependent on the outcome of the research of neuroscience, a science equally perplexed, on the question.

The quintessence of this explanatory helplessness is quite concretely on display in Oegstgeest, Netherlands. In Oegstgeest is located a several story museum, the Corpus, with walls and halls modeled in fiberglass to resemble the inside of the human body, giving visitors the sensation of being shrunk to the tiny scale of the adventurers in the film, "Fantastic Voyage." Once the visitors ascend past some dissolving cheese making its way down a hallway-sized intestinal system, they arrive eventually at the summit – the brain. Here they take seats around a cluster of display panels built atop model neurons. The neurons project images – perhaps our coffee cup on the table or the half-eaten cheese – onto a larger screen at the top of the domed space, to give an impression of how consciousness works. Unsaid is the fact that no one has a clue how neurons "project" anything in the way of an image, and if the neurons did so, to what screen? The top of the skull? And just "who" is in the brain's theater looking at the screen on the top of the skull? Now we would need to explain how that "who" sees the image on the screen. So, we would have to have another Corpus, with control room, with neurons, with screen, to explain how the first image is seen. It is what is termed an "infinite regress," for we never get out of the rather expensive need for building another Corpus within the previous Corpus.

This is the problem as it stands today. This is the problem as it stood in 1970. This is the problem as it stood when Bergson wrote in 1896. One does not need the existence of computers or robots to see this. In my graduate student days and in my doctoral thesis, I liked to call this the "coding problem." The light patterns or sound patterns of the external world are being translated to the brain's own form of "code." The external world is "encoded" in the form of these neural firing patterns. This encoding, resulting in the strange, dark "internal world" of the brain, is what I termed the standard model of perception (Figure 1.2). But now the fun begins. How, we can ask, can a code, which is supposed to stand in for something known, i.e., for the external world, itself be the means by which the external world is known? Three dots (a code), "...", can stand for an "S" in Morse code, the number 3, the three blind mice, or, if you are novelist Dan Brown, perhaps for Da Vinci's nose. How is the domain of the mapping (from code to world) specified? How is the code unfolded into a cup, coffee and a stirring spoon? That is, how is a code decoded or unfolded into the external world without already knowing what the external world looks like?

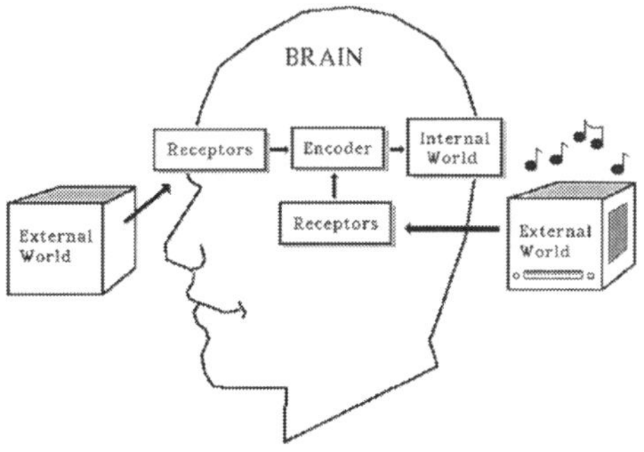

Figure 1.2 The "Standard Model" of Perception

It is not that this problem was unknown in the 1970's. Roger Sperry, the great neuro-psychologist, pointed this out in 1952, saying:

> In short, current brain theory encourages us to correlate our subjective experience [the steaming coffee cup, the singing violin concerto, the feel of rough sand paper] with the activity of relatively homogeneous nerve-cell units conducting essentially homogeneous impulses through roughly homogeneous cerebral tissue. To match the multiple dimensions of mental experience we can only point to a *limitless variation in the spatial-temporal patterning* of nerve impulses.[6]

Neural understanding was well advanced enough at the dawn of the computer model of mind to see the problem. It is in all respects the great problem of subject and object. And the computer model of mind, to any one who wished to admit it, was only exacerbating the problem. Give the computer a camera, give it a microphone, give it a pressure sensor – all of these different sources of input must be transduced to changing patterns of binary (1 or 0) bits, e.g., 111110000 or 0101011110, etc., where the bits became on/off states of magnetic cores. No matter what the complexity of the programs the computer simulators of mind would propose for handling processes of vision or hearing or touching, all must reduce to operations on, and transformations of, these patterns of bits. At its basis, starkly, for the entire computer model of mind, lay both the coding problem and the problem of subject and object. How, from these patterns of on/off bits would the computer, as subject, see the steaming coffee cup located externally, as an object, in space? How would the computer, as subject, "know" the external object as external? For David Marr, the influential theorist of computer vision, the question would sit at the heart of his framework for computer vision, for to Marr, "…the true heart of visual perception is the inference from the structure of an image to the structure of the real world outside."[7] The "structure of the image" is just the structure in the patterns of bits. The "structure of the real world outside" just happens to be our everyday image of the external world – our image of the white cup with steam wisps and swirling brown coffee from the stirring spoon. Yes, getting from one to the other is a problem, and it is one heck of an "inference."

The problem is not just germane to computer models. It runs ubiquitously today through the numerous solutions to the "hard problem," a problem, as I noted in the Introduction, famously so coined by David Chalmers in 1995. Simply put, philosophy today asks how our experience has *qualities*, for example, the "redness" of the sunset, the "rustling" of the leaves, the "singing" sound of a violin. How, Chalmers asked, after you have described your computer architecture and your software programs, or your neural networks, have you accounted for the qualia of the perceived world? This "qualia" problem is simply an aspect of the general problem of accounting for the image of the external world. The coffee cup (and its image) has qualities – it is white, the coffee is brown, the spoon is silvery – but the cup has qualities of *form* as well. As we shall see, the philosophers seem to have convinced themselves that form is not "qualia," that the form of the cup is easily "computable" and thus not subject to the hard problem. This, we shall see, is dead wrong, but just for now, it is not hard to see that if your computations have encoded form as 1's and 0's in on/off states of a computer's memory, you still have the coding problem. How is this code now unfolded as the cup? It takes great feats of magical thought to convince oneself that an array of rapidly changing bits is now, marvelously, the same thing as the image of the coffee cup.

The coding problem now goes under the "problem of representation," i.e., how does the brain "represent" the world and its coffee cups in, yes, the brain's neural codes. I prefer using the "coding" formulation for it is a more basic statement of the problem and due to the fact that the confusion over the nature of this problem is endemic. If I listed one, I could list twenty-five "hard problem" solutions, each with an intrinsic coding problem, utterly unacknowledged. We cannot take the neural-encoded information, apply an "integrating magnetic field" and claim we have explained the image of the coffee-cup-being-stirred when we cannot begin to explain how a magnetic field can unfold a "digital-neural" code. Yet this has been claimed.[8] We cannot expect RoboMary, a theoretical robot who does not perceive color, to overcome this lack simply by "self-programming" the range of "color codes" in her "color registers." The coding problem, woefully, yet remains.[9] Or, though we might propose a nice model of the brain as employing quantum processes, quantum decoherence, etc. – this says not one thing to the coding problem. You now have only succeeded in coding the world in quantum states, i.e., a quantum code. There are many more such solutions; the problem should be clear.

This statement of the "coding problem" anticipates my story, but only slightly. It did not take me long to come to it, even before my official introduction to the computer models of mind. But nevertheless, all of the thoughts above do not begin to solve Bergson's "koan." Why the relation of subject and object in terms of *time*, not space?

Koan Number Two

I threw myself into *Matter and Memory*. I read it forwards, backwards, inside to out in each direction. Bergson had spent the previous five years studying the cases of, and theories on, the loss of memories. This included amnesias which result from sort of injury or blow to the brain, or "aphasias" where some class or form of memory is lost due to a lesion in some area of the brain, perhaps, as an example, affecting the ability to recognize objects or people in vision or the meaning of words. Bergson saw a profoundly key question. It is a question to which philosophy is oblivious today. Bergson felt that whether or not our past experiences are actually stored in the brain is key in deciding whether or how the brain "generates" consciousness. In examining the cases of memory loss, he felt the evidence in no way proved that our past experiences are stored in the brain. In fact, he described a model of mind and brain in which the brain was *not* the storehouse of experiences.

Bergson introduced his analysis in *Matter and Memory* with a remarkable theory of vision. In a key passage, he disparaged the idea that the brain could in any way develop a "photograph" of the external world, for example, an internal, photographic image of our coffee cup on the table. Indeed, as I have already discussed, modern neuroscience certainly agrees with this assessment. We see only chemical flows from neurons to neurons, only spatial and temporal patterns of neurons firing. Nowhere in the patterns of neural activity could there be anything like a "photograph" or worse, a set of photographs of the cup on the table with a spoon stirring the coffee as successive snapshots of the spoon's motion. In this context, Bergson then stated:

> But is it not obvious that the photograph, if photograph there be, *is already taken, already developed in the very heart of things and at all points in space*. No metaphysics, no physics can escape this conclusion. Build up the universe with atoms: Each of them is subject to the action, variable in quantity and quality according to the distance, exerted on it by all material atoms. Bring in Faraday's centers of force: The lines of force emitted in every direction from every center bring to bear upon each the influence of the whole material world. Call up the Leibnizian monads: Each is the mirror of the universe.[10]

What did Bergson mean by the "photograph developed in the very heart of things and at all points of space?" Another koan. On my first and many subsequent readings, this was a mystery.

His admirers and his critics alike considered his discussion of perception obscure. Nevertheless, the book had a great impact at the time. He went on to publish his famously received work on evolution in 1907 (*Creative Evolution*). His fame became such that the papers debated whether he should give his university lectures at the Paris opera hall, as the ladies of society wished to hear the great Bergson. If the radical nature of his theories wasn't enough, his very popularity with the public seems to have won him some hard feelings in the academic world. The British philosopher, Bertrand Russell, attacked him bitterly. One admirer of note, however, in America was the great psychologist William James, and another friend was the great educator, Maria Montessori.

Another ally, highly influenced, whose work and similarities to Bergson I must neglect, was the great British philosopher Alfred North Whitehead. Yet others influenced by Bergson's freshness of thought included the philosophers Martin Heidegger and Gabriel Marcel. By 1914, to avoid further embarrassments due to his popularity, he resigned his teaching position.

He emerged again briefly in 1922. Stimulated and disturbed by the supposed implications of the physicist Paul Langevin's "twin paradox" (where a twin leaves the earth on a rocket and later returns having aged less than his brother who stayed behind), Bergson published an analysis of relativity, *Duration and Simultaneity*, and also in 1922, he met with Einstein for a discussion on the interpretation of the theory of relativity and the twin paradox, an issue at that time quite hot. Einstein and the world of physics were unmoved by Bergson's arguments, but this is a story I will explore later; I will only say here that this issue lies at the root of certain problems in physics to this day, to include the continued failure to reconcile relativity with quantum mechanics. By 1924 his health began to decline, though in 1932, concentrating on fields beyond science, he published *The Two Sources of Morality and Religion*, a work in which he saw the true source of religion in the dynamic, creative, mystical forces in history, and not within the conservative, traditional forces which tend to defend established systems. Virtually secluded due to his health, his isolation coincided with a general loss of interest in his philosophy (but a loss of interest in the Whiteheads and Heideggers as well). The age was to be dominated by other forms of thought. Bergson died in 1941 of pneumonia, in Nazi-occupied Paris, contracted after waiting for hours in a line (having refused all privileges) to declare himself as a Jew. By that time, relatively little academic attention was now paid to his work, particularly in psychology. Queued on the horizon was the behaviorist model of mind of B. F. Skinner with his rats and Skinner boxes. Next would come the great model of my generation – the concept that the brain is a computer.

One critic, reviewing Bergson's career and work in retrospect, would state:

> The best explanation for Bergson's impressive failure as a scientific theoretician is the same as that for his failure to succeed as a metaphysician: he was not sufficiently conversant with the outlook and problems of mathematics and physics.[11]

But this was not the view of de Broglie, a founder of quantum wave mechanics, and other physicists, who realized that Bergson had foreseen the impending crisis in the foundations of classical physics.[12] Rather, there has been an impressive failure to understand Bergson.

Enter the Computer

If I were to pick a year when the computer model of mind and brain swept into the University of Minnesota, I would say 1972. In this year the Carnegie-Mellon computer scientists, Alan Newell and Herbert Simon, published *Human Problem Solving.* My primary advisor in educational psychology, Paul E. Johnson, held a seminar on just this book, and the graduate students in the class, including me, poured over its chapters and discussed it. To Newell and Simon, the brain is just another form of computer. Its essence is to *manipulate symbols*, just like the symbols – the little a's and b's and c's – that are moved around in an algebraic equation using the proper rules:

> It can be seen that this approach makes no assumptions that the hardware of computers and brains are similar, beyond the assumption that both are general purpose symbol-manipulating devices, and that the computer can be programmed to execute elementary information processes functionally quite like those executed by the brain.[13]

In the book, Newell and Simon described a computer program, General Problem Solver (GPS), which could successfully solve problems in three subjects – logic theorem proving, chess, and cryptarithmetic. An example of the latter type of problem is, DONALD + GERALD = ROBERT, where one must find a unique number value (0-9) for each letter, where we are given initially only that the letter D has the value 5. Real human subjects were recorded, with their ongoing comments, as they solved these problems – a verbally expressed thought-trail termed a "protocol." The protocols of the human subjects and the steps they took towards a solution were "compared" with the "protocol" of the steps the computer took, and the authors claimed reasonable similarities. That the two "protocols" with their rather vague similarities might represent two completely different underlying processes or forms of thinking "devices" wasn't much worried about. This very loose form of "scientific" comparison still carries on today, both with this form of AI, with the neural network models, and ultimately with process models of building software that software companies, like the software company I once worked for, came to advocate.

In essence, the solution method of GPS was the exact form of what would eventually become "expert systems." Paul Johnson was extremely impressed. Along with his psychology doctorate, he had a masters degree in physics, and as his research assistant, I had been helping him do experiments on physics students, attempting to model the structure of their knowledge of physics concepts. Newell and Simon gave a new, concrete approach to modeling knowledge. I even played for awhile with constructing one of these "expert systems" to solve elementary physics problems. The program would be given as input a language-based statement of the problem, for example: "The truck traveled for one hour at 10 miles/hour. How far did it go?" But handling all the possible variants of even this simple problem, I soon saw, was a lost cause. This requires knowledge of the physical world, in fact, a form of knowledge that computer modelers ultimately began to call "commonsense knowledge" and which, for deeper reasons we shall see, they have all but given up on today.

Other works like Newell and Simon's were appearing or had already, for example, MIT theorist Marvin Minsky's collection of computer programs which solved problems ranging from geometry proofs to rudimentary "language comprehension." B. F. Skinner, his rats and his Skinner boxes were being swept aside. The new model of the brain, leaping into the pent up vacuum created by Skinner, for Skinner had no model in actuality, was that the brain was in essence a digital computer – a symbol manipulation device. We only needed to discover the cool software programs necessary and we would soon have language understanding programs, programs for vision, for mathematics, for chess, and of course, walking, perceiving robots. Secretaries would take dictations from their boss in English and translate them to fluent Japanese. The computer now presented a very concrete tool to implement what was in fact a deep, long held metaphysical position on the nature of mind, a position that went all the way back, as Phaedrus saw, to Plato. The achievement of all this, given we now had an absolute lock on a device to implement this long held, implicit conception of the nature of mind and intelligence, was thought by its proponents to be very, very imminent, i.e., it should have happened thirty years ago.

Enter Bob Shaw

In this year, I attended a seminar on this new subject led by Dr. Robert E. Shaw. Bob Shaw was my doctoral thesis advisor. He was a student of James J. Gibson, one of the foremost theorists in perception at the time, yes, in the problem of vision. Gibson's theory is still very influential and will play a large role in the discussion to come. Bob would later found *The Journal of Ecological Psychology – the* academic journal for the Gibson school of perception theory. At Cornell (where Gibson taught), Bob had been a post-doc in abstract automata theory for two years. Automata theory is the abstract, theoretical basis of all computing. Bob Shaw was well armed for the subject.

In the seminar, largely keying off Feigenbaum's book, *Computers and Thought*, we went through a large array of computer models at the time which dealt with some aspect of mind, be it logic problem solving, language, vision, whatever. Each was found wanting. The "want" was not a matter of persnickety detail; it was on a general principle: the computer modelers were always giving away the problem. The brain could not possibly be pre-programmed with the advantages the computer modelers were giving themselves.

What do I mean by "giving away the problem?" Consider a problem solved by Newell and Simon's GPS termed the "monkey and the bananas" problem. The problem states that the monkey is in a room with a box, the bananas hanging from the ceiling. The monkey's problem is to grasp the bananas which are out of reach – a problem solvable by moving the box beneath the bananas and climbing up. GPS, as always, follows a "means-ends" analysis. The concept behind means-ends analysis is simple. If I want to go to the grocery store (my goal), and currently I am in my house (my current state), there is a series of "sub-goals" I need to achieve.

- To get to the store, I have to drive the car.
- To get to the car, I have to get to the garage.
- To get to the garage, I have to get out of my chair.
- To turn on the ignition, I have to put in the key.
- To put in the key, I have to have the key in my hand.
- And so on.

So we move to a specified goal by achieving (through various "means" or actions) a series of sub-goals. In the monkey's case the goal is "bananas in hand" and a series of sub-goals are given by "differences." The relevant differences (defined to the program) are the monkey's place (D1), the box's place (D2), the contents of the monkey's hand (D3), and the differences are furthermore ordered in degree of difficulty from D3 to D1. This is what is termed an *object language* – a language or symbol set defining the problem environment. The operators given the program are "climb," "walk," "move box," and "get bananas." The program must discover the proper sequence of operators required to reduce the differences successively to the point where the bananas are in the monkey's hand. This is in essence simply creating a *proof*, namely, a proof that we can get from A to B. It is a proof procedure carried out within a fixed theory of the monkey's world.

The object language, then, defines a *frame of fixed features* in which the solution is carried out. But, as GPS did, giving the monkey "move box" as an operator, i.e., the *mobility* of the box, is equivalent to giving away – or better, giving away the discernment of – a fundamental feature and thus the solution path to the problem. Given this "mobility" or "move box" as an operator, it is only a matter of solving for the right sequence to apply the operators to resolve the "differences" and thus get a meal of bananas. For Gutenberg, the inventor of the printing press, it was quite a different story. In contemplating how to produce books in mass, the "mobility" of type as an intrinsic feature or operator in the problem of creating a printing device was not at all apparent. This emerged only over *analogy*:

> ...When you apply to the vellum of paper the seal of your community, everything has been said, everything is done, everything is there. Do you not see that you can repeat as many times as necessary the seal covered with signs and characters? ...One must strike, cast, make a form like the seal of your community, a mould such as that used for casting your pewter cups, letters in relief like those on your coins, and the punch for producing them like your foot when it multiplies its print. There is the Bible... The letters are moveable. *The mobility of the letters is the true treasure which I have been searching for along unknown roads*. (Letters of Gutenberg to Frere Cordilier)[14]

Gutenberg was *defining* the features of the world in which he would solve the problem. He was *creating* the object language. The features were being defined through analogy – he was seeing analogies to seals, to moulds, to stamping coins. As we start as infants, no less later as inventors, the brain is initially confronted with a world in which it must define the features. It is the *problem of the definition of features* that must be solved. The computer modelers were simply ignoring this, surreptitiously solving the problem, defining features (the object language) and then creating a logic system in which the computer could do that which it is capable of doing – manipulating the nicely pre-defined features as symbols.

The computer modelers, we shall see, are still trying to explain how the brain does *analogies* in this very same way – by defining fixed features of the world, and then creating a proof procedure which "finds the analogy." This is diametrically opposite to what is the true case. Analogy is the thorn in the side of the computer model of mind. It was a thorn in the side of Plato, the progenitor of the metaphysical framework which spawned the computer model of mind. The complete failure to explain analogy comes from this deep source of inadequacy – the computer model of mind is incapable of supporting the method by which the brain defines its world.

Near the end of the seminar, Bob gave a talk discussing the possibility of another approach to a theory of mind. He began describing holography. As holography is essential here, we need to spend a moment on what Shaw had to describe.

Holography

The "hologram" was discovered by the British-Hungarian scientist Dennis Gabor in 1947, though its full potential waited it seems, on the birth of the laser, 1963. Holography is defined as *the process of wave front reconstruction*. In considering one of the several methods of constructing a hologram, the principles we require for understanding the process are simple. Figures 1.3 and 1.4 demonstrate both the process of construction and the nature of a hologram. It can be seen from Figure 1.3 that a hologram is essentially the record of the interference pattern of two light waves. We can imagine the interference as though we had two sets of water waves coming together from two different directions. Where the crests of the water waves meet, the waves will reinforce each other, making a yet higher, more intense wave. This is termed "constructive interference" (++). Where the troughs of two waves meet, we get a wave with yet a lower trough. This is also "constructive interference" but noted (–). Where the crest of one wave meets the trough of another, they cancel out, leaving "flat" water. This is "destructive interference" or (+-). In terms of light, these cases become more intense light areas on the photographic emulsion or plate (crest meets crest), or less intense areas (trough meets trough), or just neutral (trough meets crest). This pattern of wave interaction across the area where the two wave sets meet is our interference pattern.

One of these two waves is termed the *reference* wave. It is usually emitted from a source such as a laser (as designated in Figure 1.3) which provides our very "coherent" form of light. Coherence refers to the purity of frequency. A light wave containing only a

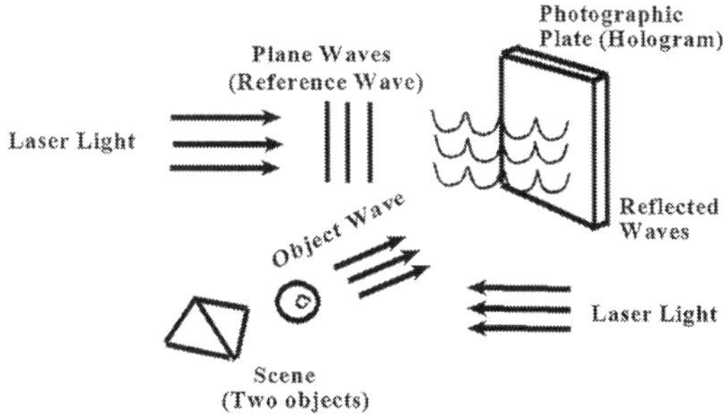

Figure 1.3 Holographic Construction

single frequency of light is perfectly coherent. The other wave, the *object* wave, arises from light reflected off the object of which we intend to make a hologram.

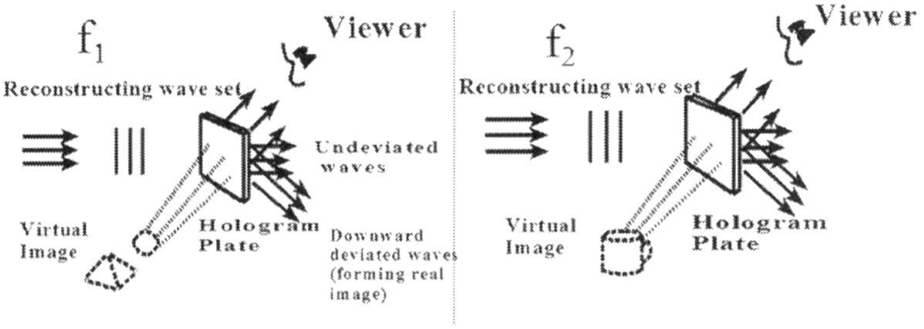

Figure 1.4. Holographic reconstruction. The reconstructive wave, modulated to frequency 1, reconstructs the stored wave front (image) of a pyramid/ball. The reconstructive wave, modulated to frequency 2, now reconstructs the wave front of the cup.

Once we have recorded the interference pattern by exposing the plate to the two waves, then we can beam another wave through the plate. This wave, beamed through at a later time, is termed a *reconstructive* wave. It is a wave of the *same frequency* as the reference wave. When it is beamed through the photographic emulsion containing the interference pattern of the two waves, the original wave front is reconstructed (Figure 1.4, left). The reconstructive wave is diffracted as it passes through the plate, analogous to that which happens to water waves as they flow through a line of barriers and bend around them, though in this case the "barriers" are the points of constructive (++) or (–) and destructive (+-) interference. Figure 1.4 shows that one set of waves travels downwards while yet another set passes right through with no deviation. A third set travels upwards in the same direction that the complex object wave in Figure 1.3 would have been moving. A viewer in the path of these upward waves then believes himself to see the source which generated the wave-set located in depth behind the hologram and in three dimensions. We might say then that the upward traveling waves "specify" the nature of their source of origination, namely a pyramid form with a globe in front of it. These waves specify what is termed the *virtual* image.

There are two major properties of holograms we should note. Firstly, we can consider each point of an illuminated object as giving rise to a spherical wave which spreads over the entire hologram plate. As the reflected light bounced off the ball, each point of the ball, for example, gave rise to little spherical wave that expanded as it traveled towards the hologram plate, ultimately covering the entire plate. Thus we can consider the information for each point on the ball to be spread over the entire hologram. This implies, conversely, that *the information for the entire object is found at any point in the hologram.* In fact, we can take a small corner or "window" of the hologram of the pyramid/ball scene in Figure 1.3 and reconstruct the image (wave front) of the entire scene with a reconstructive wave. In principle, any point of the hologram carries sufficient information to reconstruct the whole scene.

Secondly, it is possible to record a multiplicity of wave fronts on the same holographic plate. We can do this by changing the *frequency value* of the reference wave. Thus we could make a hologram of a pyramid and ball, a cup, a toy truck, and a candle successively, using reference waves of frequency f_1, f_2, f_3, and f_4 respectively. By *modulating* the reconstructive wave, i.e., changing its frequency appropriately from frequency f_1 through f_4, we could reconstruct the successive wave fronts which originated from each object (Figure 1.4, left/right). The degree of resolution of image separation is a function of the coherence of the reference and reconstructive waves. The more finely we can modulate these waves to a single frequency, the more distinctly separate and clear will be the reconstructed wave fronts. If, however, we were to illuminate the plate with a diffuse, non-coherent wave, we would reconstruct a composite image of all the recorded scenes.

Intrigued by these memory-like capabilities, the neuroscientist, Karl Pribram, had made arguments already in 1971 (*Languages of the Brain*) that the brain itself is a hologram. He postulated that memories are stored in the brain in holographic form, and thought too that some form of holographic construction might account for the external image of the world, as though, somehow, a wave is pulsing through the hologram of the brain, and reconstructing an image of the external coffee cup. The difficulty here is that if the light-wave information from the table and its coffee cup strikes the retina and is changed to a neural wave form and stored in the brain, then we need a light wave serving as the reconstructive wave passing through the brain to get the information "back outside" again in order to reconstruct the wave fronts of the table and cup. Even if such a process could work, and it can't, there is no such light wave, passing through the brain, to be found.

Pribram's conjecture was mostly ignored in the computer-entranced circles of cognitive psychology, but it was in the background of Shaw's thoughts. I should note that Karl came to Minnesota to give a lecture around this time and had conversations with Bob. Bob's concepts, as I'm about to describe, had to subsequently lurk in Karl's mind.

The Brain as Reconstructive Wave

Concluding his discussion of holography, Bob proceeded to drop his bomb. Isn't it possible, he wondered, that the universe is itself a hologram? As such, the hologram would not so much be "in the head" as Pribram was speculating, but rather, the head would be "in the hologram."

As Shaw's discourse progressed, Bergson's second koan became clear. Bergson, writing in 1896, had already anticipated the abstract essence of Gabor's discovery. This is what he meant by "*the photograph...developed in the very heart of things and at all points of space*." At every "point in space" in a hologram is the information for the whole. It was no wonder that Bergson's contemporaries found his theory elusive in 1896. Bergson visualized the universal field of matter as a vast field of "real actions." Any given "object" acts upon all other objects in the field, and is in turn acted upon by all other objects. It is in fact obliged:

> ...to transmit the whole of what it receives, to oppose every action with an equal and contrary reaction, to be, in short, merely the road by which pass, in every direction the modifications, or what can be termed *real actions* propagated throughout the immensity of the entire universe.[15]

What Bergson was arguing, in his vision of a vast field of "real actions," was that, indeed, the matter-field – the universe – is holographic, a vast, ever changing, dynamic interference pattern, in fact, given its transformation over time, a four-dimensional hologram.[16] Indeed, in 1980, the physicist, David Bohm, would first introduce the idea in physics that the universe (what I like to term the field of matter or the "matter-field") is holographic. He termed it the "holofield." Since then, physics has come to rather routinely view the matter-field as forming a vast, dynamic interference pattern, i.e., a hologram.[17]

Let us consider an intuitive model of this hologram on a universal scale. Visualize the universe as a vast fluid. Then imagine each little point or tiny fluid-particle in the fluid as a "vibrator" giving off a spherical, rippling wave that expands throughout the entire fluid (Figure 1.5). These waves are Bergson's "real actions" propagating throughout this universal fluid-sea. Each little vibrating point influences all other points in the fluid with its spherical, spreading waves. Simultaneously, each individual point is being influenced by the spherical waves from every other point. The "state" of each little vibrating point reflects the whole. As the Greek, mystical philosopher, Plotinus once said of his vision of reality, "There each part always proceeds from the whole, and is at the same time each part and the whole." This system of waves expanding everywhere, from every point, becomes a vast, dynamic, ever changing interference pattern of waves – a holographic field. To capture this dynamism, Bohm (*The Implicate Order*) would later call it the "holomovement."[18]

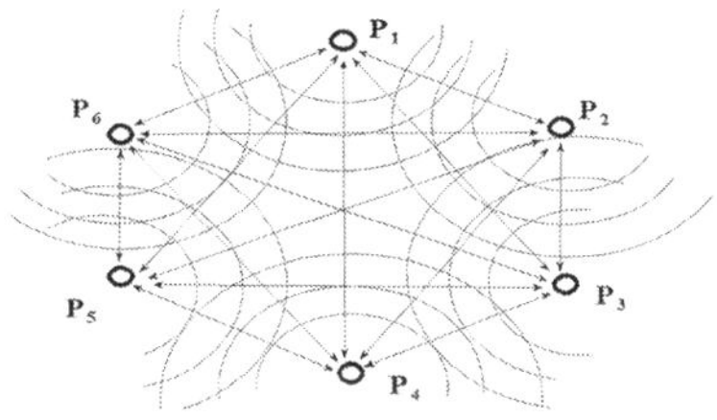

Figure 1.5. A mini-universe of six vibrating points in a fluid. Each point influences all the others with its radiating, spherical wave

After Shaw's talk, I wandered through the rooms and corridors of the psychology building in a strange state. I was within a vast Hologram. The walls, the floors, the fine grain of the wooden doors seemed to gleam with a crystalline glow, pulsating just a little. But there were many theoretical difficulties to face. If the matter-field is holographic, how does this help solve the problem of vision? Bergson, I knew, had gone immediately beyond Pribram's difficulty (his 1971 version) where there is some form of reconstructive wave passing internally through the brain – a wave impossible to find. Following hard upon the passage concerning the "photograph... at all points in space," Bergson stated the following:

> Only if when we consider any other given place in the universe we can regard the action of all matter as passing through it without resistance and without loss, and the photograph of the whole as translucent: Here there is wanting behind the plate the black screen on which the image could be shown. Our "zones of indetermination" [organisms] play in some sort the part of that screen. They add nothing to what is there; they effect merely this: *That the real action passes through, the virtual action remains.*[19]

In essence, Bergson was arguing, with Shaw, that the *brain is the reconstructive wave*. The brain, seen as itself a very concrete wave, is "passing through" this holographic field, and specifying the external image of a portion of this field – say, our coffee cup. Just as we saw that a reconstructive wave of a certain frequency, passing through the interference pattern on a hologram plate, decodes or picks out a certain wave front to specify, say, a pyramid/ball vs. a cup, so the portion of the matter-field that the brain-supported reconstructive wave picks out is a portion related to the body's action. Thus, perception, as Bergson was saying, is *virtual action*. We are seeing how we can act.

The brain, then, is a *decoder*. Rather than encoding the external world within in an internal neural code that we now are at a loss to explain how to unpack, the brain as a reconstructive wave passing through the holographic matter-field unpacks an action-relevant set of the mass of information (or real actions) in this field, specifying it as an image, be it a cup, a buzzing fly or a stirring spoon.

The concept that the brain might be a very concrete form of wave may seem surprising. Nevertheless, there are many things in current neuroscience that point to this possibility. The visual processing areas of the cortex, V1, V2, V3, V4 and V5 are all interconnected and feed information forwards and backwards to each other, creating a system "resonating" with feedback. V1 for example, which processes the initial flow of information received by the eye, feeds information to V5 which processes information related to motion, while V5 in turn sends information back to V1, modulating V1's own processing of incoming information. All these areas connect to the motor areas of the brain which control action, and the motor areas in turn feedback to the visual areas, modulating, in turn, visual processing. There are other wave-like "oscillations" in cells or areas of the brain (for example 40 Hz gamma oscillations) that are believed now to be phase-locked in a form of synchrony across the brain.

These oscillations and the resonating feedback are *abstract* waves. For our reconstructive wave, we need a very *concrete* wave, as concrete as the field of force generated by an AC motor. No, I'm not saying the brain *is* an actual AC motor; I am compelled to use this comparison to emphasize the need that exists for describing a very concrete dynamics for the brain.[20] The fact is, all the "computations" that computer modelers believe are occurring across the resonating feedback among the various processing areas noted above must be part of, in their ultimate effect, a very concrete, reconstructive wave.

This makes the brain, obviously, a very different type of "device," very different from a computer or from a neural network being used for computations. The computer or network is only interested in achieving its abstract computations – its symbol manipulations as Newell and Simon insisted. As long as these can be achieved, whether on the beads of an abacus, on a computer with transistors, on a computer with vacuum tubes, whatever, achieving this symbol manipulation is all that is important. But the brain is in truth creating a very concrete waveform within the holographic field, for achieving modulations within that field. It is this *real dynamics* that is all important, as important as the dynamics that must be achieved in an AC motor. One does not build an AC motor out of plastics, or rubber bands, or mud, or beads. There is a real dynamics that must be achieved to create the electromagnetic fields. This is equally true of the brain.

The Information for Specifying Events

A reconstructive wave passing through a hologram specifies a certain image or wave front depending on its frequency pattern – on how it is modulated. It might specify a pyramid-and-ball, or it might specify the cup (as in Figure 1.4). What drives the particular pattern of modulation of the brain's reconstructive wave? This is where we need to bring in Shaw's mentor, J. J. Gibson.

Gibson (*The Perception of the Visual World*, 1950) saw that the surrounding environment contained a great deal of mathematical information useful to the brain. The information specifies the depth of objects, the form of objects, and how the body can act upon them. One such piece of information is what he termed the *optical flow field* (Figure 1.6). We see a flowing field like this routinely when we drive down the road in our car. The arrows in the diagram are "velocity vectors." The longer the arrow, the greater the speed of that portion of the field flowing by. The field is flowing by at the greatest velocity near the eye of the observer (or the driver of the car). There is a still-point at the origin of the field, up in the distance, called the "point of optical expansion." It is near the beginning of the mountains in the figure. This entire set or array of velocity vectors (arrows) flowing by is called a "gradient," as a gradient changes gradually. In this case it is gradually changing in terms of the values of velocities, from zero to ever larger. This gradient happens to have a precise mathematical ratio – information useful to the brain. The velocity value of each vector is inversely proportional to the square of the distance from the eye.

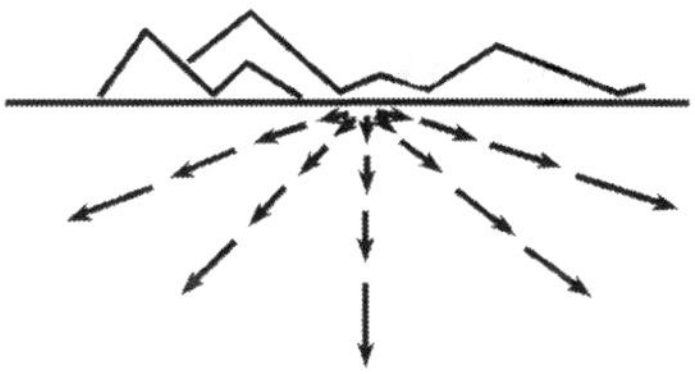

Figure 1.6. Optical flow field with its gradient of velocity vectors.

Gibson, it seems, was struck by these fields as a young boy while riding on the cabooses of freight trains, on which it appears, he was in the habit of catching a ride. Of course, while riding a caboose, the fields are flowing away from the observer, and thus contracting towards a point, not outwards and away from a point. The actual surface of such a flowing field is also highly structured. We might be walking across a field of small rocks near the mountains, or a beach with its grains of sand, or a prairie with its grass, or across our kitchen floor with its tiles. These surfaces have what Gibson termed a *texture gradient* (Figure 1.7). Imagine the surface of Figure 1.7 as a rocky or gravel surface. The little circles (rocks in this case) are the "texture elements." The size of these elements and their horizontal separation changes in perfect mathematical proportion with the distance from the eye. The vertical separation between each element decreases in proportion to the square of the distance. These texture gradients are ubiquitous – floors, beaches, lake surfaces, tiles, etc. Turn the gradients upside down – you see them as ceilings or the bottom of clouds.

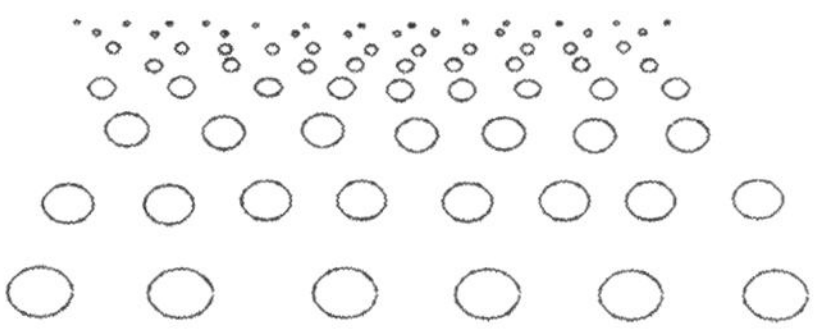

Figure 1.7. Texture density gradient (after Gibson, 1950)

The mathematical relations or information of these gradients are preserved on the retina by the natural projection into the eye (Figure 1.8). Note in Figure 1.8 how the distances along the ground line (G_1G_2) are preserved when projected into the retina. The distance relations on the ground (W to X, X to Y, Y to Z) are preserved on the retina, but in reverse order. In fact, before Gibson, the problem of how we determine the distance of some object had been rather mysterious. Bishop Berkeley, an influential philosopher on the subject in the 1700's, had framed the problem in terms of the line ABCD in Figure 1.8. In this case nothing varies on the retina as the eye moves forward, for each point A, B, C, or D projects to exactly the same retinal point!

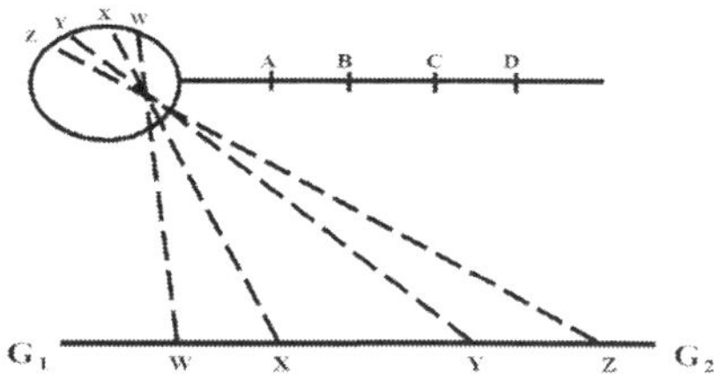

Figure 1.8. The "Ground"

Gibson argued that these gradients naturally "specify" distance, in fact, specifying the stretching surface of the external environment. The information is received on the retina and processed by the visual areas of the cortex. We discussed how the visual areas V1 through V5 and the motor areas all "resonate" with feedback from each other. This is a post-Gibson discovery of neuroscience, yet at the time, Gibson felt the brain simply "resonated" to this information. There is no picture or image or photograph developed in the brain. The brain's state "specifies" or is "specific to" the external environment. Note the use of the term "specifies." We have seen it used in the holographic context, for the reconstructive wave "specifies" the image. Gibson never explained the origin of the image of the environment. Effectively, he was presuming Bergson's holographic field. It is Gibson's "resonant" information that is modulating the reconstructive wave.

Let us place an object on our gradient. We'll imagine a coffee cup resting on the surface of a tiled table or patterned table cloth (Figure 1.9). In Figure 1.9, I want you to imagine that we have the *same* cup moving to two different positions – from back to front. The forward, larger view of the cup only *occludes* (or covers) two rows of (largish) texture units. The rear, smaller view of the cup occludes several layers or rows of smaller texture units. In fact, as the rear cup is moved to the forward position, there is preserved a constant, inverse ratio of the size of the cup to the number of texture units occluded. As the size (height) of the cup grows, the number of layers of units it hides decreases.[21]

Figure 1.9. Texture density gradient (or tabletop) with a cup in two different positions.

This is what is termed an *invariance law*. The ratio of cup height to texture units is an invariant – a ratio that does not change. It is this invariant that specifies the *size constancy* of the cup. The event of a cup moving towards us from one position to another is always experienced as a *cup of the same size* moving across the table – despite the growing size of the cup per se on the retina as the cup approaches. It is this invariant ratio that the brain is picking up that enables it, as a reconstructive wave, to "specify" that there is a cup of constant size moving across the table.

My first introduction to invariance laws was through a doctoral thesis, "On the Structure of Laws of Invariance," by Dr. Larry Victor. Larry had a PhD in physics and just before my time had written his thesis for his PhD in educational psychology under my advisor, Paul E. Johnson. In the introduction to the thesis, he also acknowledged one, "Professor Cannabis," as a source of inspiration. As a physicist, he appreciated the significance of invariance laws, as these are precisely what physics attempts to discern in the physical world. Larry was trying to describe a general framework in the education system which would support a child's perception of the laws of invariance in any subject. Perhaps due to my lack of acquaintance with Professor Cannabis, the thesis was obscure to me, but we shall see that it all eventually became very concrete in the context of mathematics education.

Invariance laws were fundamental for Gibson. We shall keep meeting them. The proportional gradients of flow fields and of texture unit distributions are also invariance laws. Physics prides itself on discovering the invariance laws of the universe. Newton's F=ma and Einstein's $E=mc^2$ are invariance laws. The discovery of invariance laws is the essential endeavor of science, and in this effort, physicists only model themselves after the method of the brain in perception.[22]

As we go on, we shall see that any event, for example, simply stirring coffee with a spoon, is structured or defined by numerous invariance laws. The event has what I shall call an *invariance structure*. This will become even more pronounced as we start considering the time-extension of an event. I have already stressed that it is crucially important to build the foundation of the real, concrete dynamics of the brain, to include all its "levels" – quantum, chemical, neural – as critical as establishing the real, concrete, electromagnetic dynamics of an AC motor. As we begin to conceive of and model the brain, with its neural, chemical, even quantum dynamics as supporting a concrete reconstructive wave, it is all-important that our future conceptions show how this invariance structure is simultaneously being supported or incorporated. It is *utterly useless* to be offering models such as the "quantum brain interacting with the quantum level events of the Zero-Point field and (say, via Fourier transforms) projecting an image" and stopping there, while ignoring the mass of invariance information in the ecological event that the brain is actually processing and which *must*, ultimately, be mapped to any quantum level (and beyond) description. In the first place, by the principle of virtual action, as we shall further see, it is this ecological invariance structure as *information* that is being used by the action/motor areas of the brain to modulate the visual areas and select, from the totality of real actions in the quantum/holographic field, that portion that is specified as virtual action (i.e., an image of the field). It *is this invariance information at the ecological level* – and ecological invariance laws are the true *information* – that we need for perception theory, memory theory and cognition.

The Picture Thus Far

What we have seen thus far is that the matter-field can be viewed as a hologram of cosmic scale. The brain in turn can be viewed as a reconstructive wave passing through this holographic matter-field, and specifying an image of the field, or better, a portion of this field. The image, say, of the cup, is *precisely where it says it is* – external, *in the field.* This is termed *direct* perception. The modulation of this reconstructive wave is driven by the mathematical (invariance) information of the external events which are occurring in this dynamically changing, external field.

I should note that when we say that the brain as a reconstructive wave is "specifying the image," this cannot be taken to mean that an image is being *projected* by the brain into the field or into somewhere – or anywhere. "Specifying" is simply shorthand for "the state of the brain is specific to" a source in the holographic field. As I noted earlier, Gibson was fond of saying that the (resonating) state of the brain, via the invariance laws in the external world such as his texture gradients, "is specific to" the external environment. He had no image arising in the brain or "generated" by the brain (or arising within the circle of Figure 1.1). The "generation" notion, a never-to-be-explained mystery, is something, as we have seen, to which *indirect* realism is inherently wedded. His "specific to" is to be taken in the same sense that a reconstructive wave, passing through the hologram plate, is "specific to" a certain source, e.g., the reflected (object) wave that came from the pyramid/ball. The "image" is already inherent in the hologram plate, as each of its points carries the information for the whole. The reconstructive wave is simply selecting out, and specific to, a certain wave-front source.

The brain, then, is specific to a "source" in the holographic matter-field, and more precisely, as we will later see, it is specific to a source that is in fact a *past* motion of this field taken at one of many possible scales of time. Further, given the very nature of the motion of the field, it can only be an optimal (or probablistic) specification. This latter fact will destroy any possibility of this direct realism being construed as a "naïve" realism. The reason that there is no need for the brain to generate anything, to generate an image, is that within the holographic field – massive interference pattern that it is – all possible images are contained. The field itself is non-imageable – one cannot form an "image" of this massive interference pattern wherein each "point" reflects the whole – Bergson's "photograph developed in the very heart of things and at every point of space." But given a principle of selection, of carving out – a process of specification, in this case via a reconstructive wave – then that process (via the brain) is specific to a subset, or as Bergson said, to a *diminution* of the whole, a diminution that by this very process *is* the image. The brain need merely be "specific to" some selected aspect of the field, to some source of one of these patterns, some original "object wave" – now an "image," and it is within this external, holographic field that the "image" resides. But there are many things we have not filled in. I will list them.

Firstly, the events of the matter-field have long since come and gone by the time the information is processed in the various areas of the brain with their "resonant" feedback. A buzzing fly moving by, for example, his wings oscillating at 200 beats/second, has not only changed position but has executed many wing beats. Practically, the brain is *always* specifying the *past.* More precisely, the reconstructive wave is specifying a past motion or transformation of the matter-field, in this case, that portion of the field which is (was) the buzzing fly. How is specification of the *past* possible? How can we be looking at a past motion of the field?

Secondly, we must ask, who is looking at this (past) image? We hope our answer is not, "a little man seated in a control room in the brain." If it *is* our answer, we can expect Bassui to rush from Nirvana, crack us soundly about the shoulders with his kyosaku and command us to redouble our efforts in zazen (Zen meditation). "Concentrate on your koan," he would say, "Who is it that sees? You are yet clueless!" Though this "little man in the control room" happens to be an image presented in PBS specials adoringly recounting the achievements of our current neuroscience of vision, it is an answer we must avoid. It is called the problem of the *homunculus* (precisely, "little man"). It is an infinite regress – for we now need, like the Oegstgeest display, a second little man inside the first little man's brain, ad infinitum. This is the implicit difficulty with any simplistic proposal that rests with an image projected externally, somehow, from the brain, whether or not it presumes that the brain is using quantum processes, whether or not it presumes the brain is embedded in the zero-point holographic field. *Who now looks at the image*? We would now need a theory of an "image processor." To put the problem equivalently, we must remove the "viewer" in our picture of holographic reconstruction (Figure 1.10). In essence, we have yet to understand how the relation of subject and object is in terms of *time*, not space.

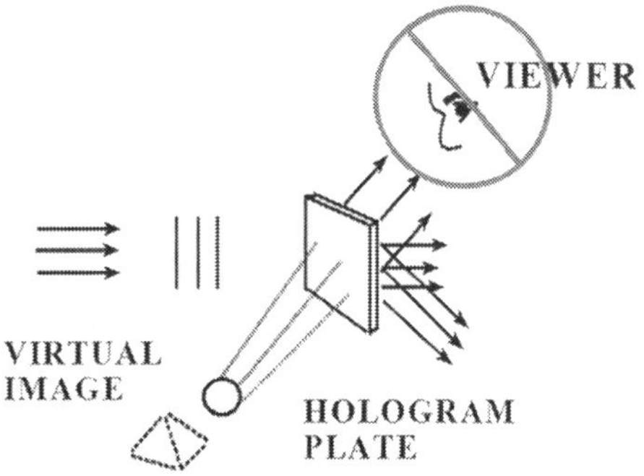

Figure 1.10. No "viewer" allowed. We must remove the viewer from our picture of holographic reconstruction.

Thirdly, we must address the "qualia" problem, the origin of *qualities* of our experience – the "blueness" of the lake, the "rustling" of the leaves. This "hard problem" is only a formulation of aspects of the problem of the origin of the image of the external world. The problem of the origin of the image of the world has been pondered since the Greeks.[23] But even this history of thought in the problem of perception and its image of the world seems to have been forgotten by modern philosophy. The image clearly has qualities of color. But the changing forms composing the image, be it "buzzing" flies, "rotating" cubes or "rustling" leaves also have unique qualities, ignored by philosophy. Ignoring the coding problem yet involved, as we have seen, forms (thus objects having form) are thought "computable" and thus capable of machine vision. This is a distortion based on an incorrect conception of time – a conception corrected long ago by Bergson. But philosophy assigned Bergson to the dustbin long before his death. He in turn was too diplomatic to take them to task as would have a Bassui.

Fourthly, we must explore the implications of Bergson's principle that perception is "virtual action." I have given this short shrift here. We have yet to see how Gibson's "invariants" are used by or "get into" the action systems of the body and brain. There are profound implications in this, implications unto the possibility of action in altered time.

It is to these questions that we will turn next.

Where We Are Going

I have swept us up into the problem of perception. What we have seen is that to explain perception, our model of the brain – which I have just begun to paint – must be vastly, profoundly different from the computer model of mind. The story is actually far worse yet for the computer models, and I will put the finishing touches on the perception aspect of things in the next chapter. What we have seen and will see even more is that perception, which is to say, our *experience*, is *not occurring solely within the brain*! If experience is not occurring solely within the brain, then it cannot be solely *stored* in the brain. Contrary to what is to all practical purposes a dogma, the brain cannot be the storehouse of experience. This means that there must be an entirely different model of memory – far different than the "storage-in-the brain" model that the computer or neural network models presume. Yet thought (or cognition) obviously employs elements of our past experience. The nature of thought too must be entirely different than that envisioned by the computer model of mind.

We, as human beings, are far from computer-like robots creating widgets, whether it be software widgets as in my own former workplace, or dress designs, or electronic switches, etc. In fact, we shall see, the computer model has absolutely no clue how we create widgets. Also in fact, evolutionary biology is similarly clueless as to how the *universe* creates widgets, be the widgets butterflies, walruses or brontosauri. We shall see that the mist of the robotic image of man that we have all been breathing in, and implicitly infusing in us views on the nature of mind and brain and man, is saturated with falsehood.

Chapter I: End Notes and References

1. Gunter, P. A. Y. (1969). *Bergson and the Evolution of Physics.* University of Tennessee Press.
2. Not famous enough, apparently, to prevent Amit Goswami from publishing a book recently (2008) entitled, exactly, *Creative Evolution*, without containing even a single reference to Bergson's great work.
3. Bergson, H. (1896/1912). *Matter and memory.* New York: Macmillan. (Originally published 1896).
4. Crooks, M. (2002). Intertheoretic identification and mind-brain reductionism. *Journal of Mind and Behavior, 23,* 193-222.
5. Hawkins, Jeff. (2004). *On Intelligence.* New York: Times Books, p. 63.
6. Sperry, R. W. (1952). Neurology and the mind-brain problem. *American Scientist. 40,* 291-312.
7. Marr, D. (1982). *Vision.* San Francisco: W. H. Freeman.
8. McFadden, J. (2002). Synchronous Firing and its influence on the brain's electromagnetic field: Evidence for an electromagnetic field theory of consciousness. *Journal of Consciousness Studies, 9,* 23-50.
9. Dennett, D. C. (2005). *Sweet dreams: Philosophical obstacles to a science of consciousness.* Cambridge, Massachusetts: MIT Press.
10. Bergson, H. (1896) *Matter and Memory*, p. 31, emphasis added.
11. Hanna, T. (Ed.), *The Bergsonian Heritage.* New York: Columbia University Press, 1962, p. 23.
12. De Broglie, L. (1947/1969). The concepts of contemporary physics and Bergson's ideas on time and motion. In P.A.Y. Gunter (Ed.), *Bergson and the Evolution of Physics,* University of Tennessee Press.
13. Newell, A., & Simon, H. (1961). Computer simulation of human thinking. The Rand Corporation, p. 2276, April 20, 1961, 9.
14. Montmasson, J. (1932). *Invention and the Unconscious.* New York: Narcourt-Brace, pp. 273-274, emphasis added)
15. Bergson, H. (1896/1912). *Matter and Memory*, p. 28.
16. Robbins, S. E. (2006). Bergson and the holographic theory. *Phenomenology and the Cognitive Sciences*, 5, 365-394.
17. Bekenstein, J. 2003. Information in the holographic universe. *Scientific American* 289(2): 58-66.
18. Robbins, S. E. (2000). Bergson, perception and Gibson. *Journal of Consciousness Studies. 7,* 23-45.
19. Bergson, H. (1896/1912). *Matter and Memory,* pp. 31-32, emphasis added.
20. It is unfortunate to have to even make this warning, but yet this (the brain is an actual AC motor) is actually how a reviewer of one of my papers took this (and rejected the paper). Such incidents hardly inspire confidence in the academic community.
21. We'll call the object's size or height on the retina, S. We'll let N = the number of texture units the object occludes. The inverse ratio is S = k/N, or SN = k. If the size (S) of the object on the retina is 8 units and the object occludes 2 texture gradient units (N),

then S x N = 8 x 2 = 16. At a further distance away, the object's size on the retina is 4 units, and 4 texture units are occluded, or again, S x N = 4 x 4 = 16.

22. Woodward, J. (2000). Explanation and invariance in the special sciences. *British Journal for the Philosophy of Science*, 51, 197-214.
Woodward, J. (2001). Law and explanation in biology: Invariance is the kind of stability that matters. *Philosophy of Science*, 68, 1-20.
Kugler, P. & Turvey, M. (1987). *Information, Natural Law, and the Self-assembly of Rhythmic Movement.* Hillsdale, NJ: Erlbaum.

23. Lombardo, T. J. (1987). *The Reciprocity of Perceiver and Environment: The Evolution of James. J. Gibson's Ecological Psychology*. Hillsdale, N.J.: Erlbaum.
Lombardo's work contains a good history of earlier thought on the problems of perception.

CHAPTER II

Time: The Mystical Perception of Stirring Spoons

I came to realize clearly that Mind is no other than mountains and rivers and the great wide earth, the sun and the moon and the stars.

- Sono-o, Zen Master, 1688-1703.

For where is the borderline between perceiving and remembering? ...Where do percepts stop and begin to be memories, or, in another way of putting it, go into storage? The facts of memory are supposed to be well understood, but these questions cannot be answered.

- J. J. Gibson[1]

The Wobbly Cube

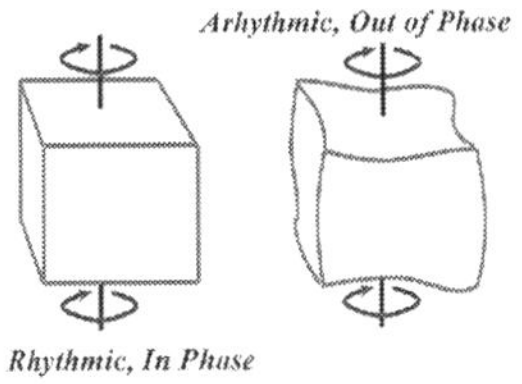

Figure 2.1. Rotating cubes, strobed in phase with, or out of phase with, the symmetry period.

How do we see the *past*? How is this even possible? I begin with a simple demonstration of the importance of this question, and therefore, of the importance of the relation of the brain to time. Bob Shaw, who deeply appreciated the problem of time, presented the demonstration to his little seminar in the 1970's. It involves a cube constructed of wire edges and rotating at a certain speed. As the cube is rotating, it is strobed periodically with a strobe light.

A cube, as it rotates, has what is termed a "symmetry period." In the case of a cube, it is a period of four, for the cube is completely carried into itself or "mapped onto" itself every 90 degree turn as it revolves. An equilateral triangle would have a period of three, being carried into itself every 120 degrees. The form of the cube is unchanged or invariant after every 90 degree turn. The form of the triangle is unchanged or invariant after every 120 degree turn. Symmetry, it should be understood, is simply another word for *invariance*.

How the strobe relates to the cube's symmetry period is critical. The strobe can be in phase with the symmetry period or at an integral multiple of the period. For example, the cube can be strobed 4 or 8 or 12 times per complete revolution (4 times/revolution = 1 x 4 or a multiple of 1, or 8 times/revolution = 2 x 4 or multiple of 2, or 12 times/revolution = 3 x 4 or a multiple of 3, and so on). If so, the cube is perceived as indeed a rigid cube in rotation (Figure 2.1). But if the strobe is out-of-phase, not at an integral multiple, for example, 9 times a complete revolution, the "cube-ness" is gone. What is perceived is a distorted, wobbly, non-rigid, plastic-like object. The standard "features" of a cube – the straight edges, the sharp vertices, the rigidity – are gone.[2]

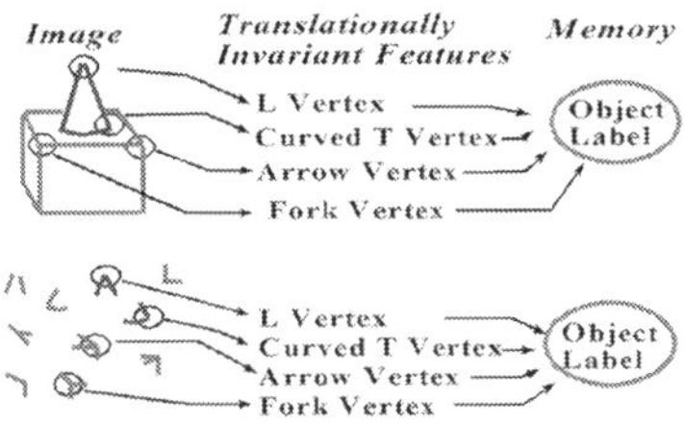

Figure 2.2. Both "objects" have the same features, and therefore would be "recognized" on the basis of a feature match.

The "features" of a cube – its edges and vertices – are normally held to be *intrinsic* to the object. An example of this thinking can be seen in a neural network model for object recognition, a model named JIM. The authors of the JIM model had noted that if the input to the brain is a set of features – edges, vertices – the features can be reassembled in many ways that still "match" (Figure 2.2). What's missing is the nice picture on the top of the puzzle box that allows the brain to put the pieces together in the right way. JIM proposed just such a picture, and it came in the form of elementary forms termed "geons."[3] Examples of a geon are a wedge, a brick, a cone. The JIM neural network linked the features to a "geon cell," preserving the spatial relations, and supporting recognition of the object.

This is a static conception of form. It collapses in the face of Shaw's demonstration. Each strobe of the cube in rotation is equivalent to a snapshot of the cube. In the snapshot should be the information for the vertices and edges of the cube. With each snapshot, the network would transmit this information to the geon cell, and the brain would perceive a cube. But Shaw's demonstration shows that there is no such information in a snapshot. As far as the brain is concerned, there is not enough information in a snapshot for a cube. This is because, we shall see, to the brain, there is no such thing as a "snapshot." The concept of a snapshot is based in a complete misconception of time.

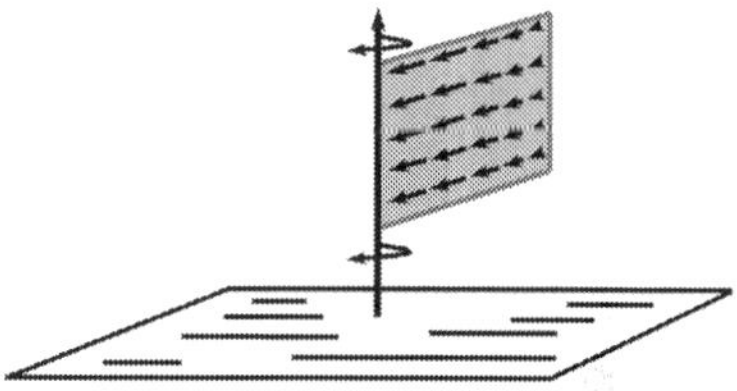

Figure 2.3. The flow as a flag rotates towards the observer.

What is so important about time to the brain, or about information over time? We return to flow fields. The flag of Figure 2.3 is rigid and stiff, and it is rotating towards us. As the flag rotates, there is a flow field over the surface of the flag. As it rotates towards us, the flow field expands (the velocity arrows get larger the closer to the eye). As it rotates away from us, the flow field contracts. A "Gibsonian" cube would be a partitioned set of these flow fields (Figure 2.4), with flow fields for each side and on the top. As a side of the cube rotates towards us, the flow field expands. As the side rotates away, the flow field contracts. The top surface of the cube is a revolving or "radial" flow field.

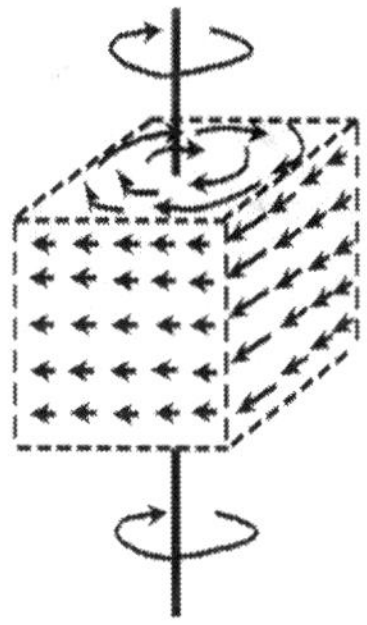

Figure 2.4. The Gibsonian Cube

What are the "edges" or "vertices" of a Gibsonian cube? They are merely sharp discontinuities at the junctures of these flows. In vision, the "features" of objects are very much creatures of time and of these flows. Even for a static, stationary cube, the eye is actually in constant motion, darting over the cube, picking up information over its own, eye-created flow.

Current perception theory relies deeply on these flowing velocity fields to explain how the brain specifies form. This reliance has its origin in yet another problem, known as the "correspondence problem," lurking in the concept of taking a series of snaphots or samples of a transforming event Suppose the cube-and-cone structure of Figure 2.2, with all its "features," is rotating. If we are taking samples of this rotation at brief intervals, in successive frames, as in a movie film, the correspondence problem is generated by considering the possibility that the visual system matches corresponding points or features in successive frames, in this example, keeping track of the identity and current whereabouts of each vertex and line of the cube-and-cone from frame to frame. For each feature, the system determines the distance traveled (d) and the time (t) between frames, and computes, the velocity of this motion (v= d/t). But this form of matching model, which must keep track of the identity of each feature from frame to frame, and particularly given the question of what constitutes a "feature," has proven to be totally intractable.

This undefeatable correspondence problem caused the gravitation to flow fields. But since these fields are moving, there are two *uncertainties* introduced, making the whole process very probabilistic, very much guesswork for the brain. The first source of uncertainty is termed the "aperture" problem. The eye has a set of little receptive fields mapping into the brain, but each of these fields is limited – an aperture – covering only a certain circular portion of a total flow field. The velocities of the flow cannot be estimated with certainty due to the limited view of each field. Figure 2.5 shows the problem. The card with the lines is moving to the right, so the card and its lines actually have a *horizontal* velocity. But when the card passes under the limited aperture, and the ends of the lines are obscured, only a downward motion of the lines is seen. The aperture forces an inherent uncertainty in measuring the actual velocity of the lines.

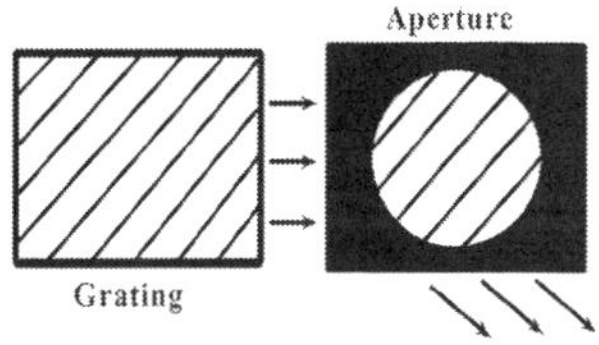

Figure 2.5. The aperture problem. The card with the grating is moving to the right, and passes beneath the card with the circular aperture. Only a downward motion of the lines (and the velocity of this motion), and not the rightward, is seen in the aperture.

The second source of uncertainty, as we shall explore later, is simply that no velocity can be *instantaneously* fixed in the first place due to the very nature of motion. The brain must do the best it can to specify form. *All* form is an optimal specification based on the best information available. Another example of this is Mussati's illusion (Figure 2.6) where we have a rotating ellipse. Each point on its perimeter can be conceived as following a path, or vector, with a certain velocity (the little arrows in the figure). At a high enough velocity of rotation, the ellipse also becomes a non-rigid, wobbly figure. As perception theorists, Weiss, Simoncelli and Adelman, argued in their foundational work on this model, even "illusions," for example, our wobbly cube, are *optimal specifications*.[4]

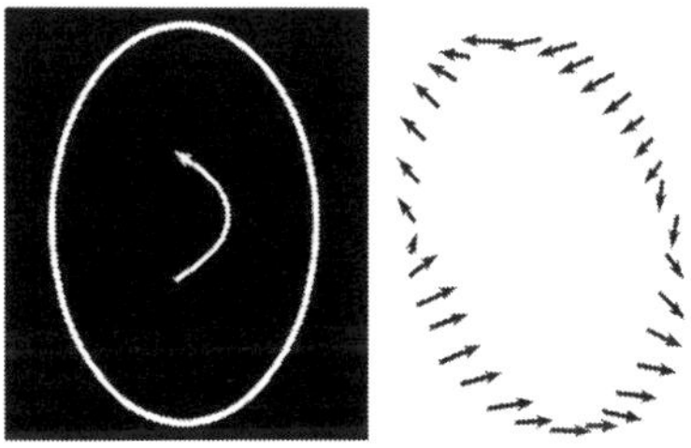

Figure 2.6. A rotating ellipse becomes a non-rigid figure at sufficient velocity of rotation. The velocity vectors (on the rim) tend to induce this non-rigidity.

In its optimal specification approach, the brain is employing probabilistic (i.e., Bayesian) *constraints* in its computations. For Weiss et al., the premier constraint is "motion is slow and smooth." This is in effect an invariance law capturing what is most often true or probable about motions in the environment. This rule/law can be expressed as a mathematical constraint incorporated within the brain's computations, and so it is in the Weiss et al. model. If the motion becomes too fast, as in an increasing speed of the rotating ellipse, and as this constraint is violated, we obtain the apparently anomalous specification of the non-rigidity of the ellipse. For the wobbly non-cube, there seems to be another constraint or invariance being violated, namely, that a rotating regular object (such as a cube) displays a regular periodicity in time. The arrhythmic strobe is effecting a violation of this constraint, leading to an "illusory" specification of a non-rigid, wobbly non-cube.

The existence of illusions has been an argument for the concept that all perceptions are "in the brain," that perception is only *indirect,* that we are not actually seeing the object but rather some internal image, mysteriously generated by the brain. This is not true. The specification of form, via the reconstructive wave of the brain, is nevertheless a specification of the past transformations of the external matter-field – *given the best available information*. The object/form is right where it says it is – in the external field. As an analogy, an imprecise reconstructive wave, comprised of components of frequency 1 and frequency 2, might "select" and specify from the information in the hologram plate a composite, blurred image of both the pyramid-ball and the cup of Figure 1.4. This is actually an optimal specification too. This does not mean that the composite image is an "illusion" just because, from a God's-eye point of view, we know that a *pure* f_1 would produce a pyramid-ball, or a *pure* f_2 would produce a cup. The composite image is also and equally legitimately available from the information in the hologram.

The eye jumps to various points in the scene (or on the cube) in movements called *saccades*. During each jump, in the transition to the next point, it takes in no information, yet the world, with its rotating cube in front of us, does not shut off and on intermittently with each saccade. How, says the argument for indirect perception, could we be still be looking at the actual cube? So, the indirect position holds, the cube must be only an "internal" image; we do not see the real cube, external to us, in space. But the brain-as-reconstructive wave does not cease during a saccade; it continues to specify the rotating cube – right where it says it is, external to the body, in space.

The Scale of Time

If flowing velocity fields are the basis for form, there is yet another factor profoundly influencing our vision of these forms. This is the scale of time. The "buzzing" fly is an index of our scale of time. Why does the buzzing fly, as we see him in our normal world, not look like a heron, slowly flapping his wings? Or perhaps like a nearly motionless fly, whose liquid crystalline biological structure we can almost see vibrating? The brain is imposing a scale of time upon the matter-field and the events within it.

The brain is conceived as a hierarchy of levels. These range from the atomic, to the chemical, to the molecular, to the neural. Each is inclusive of the other. But despite these conceptual "levels," the biological brain is a *coherent* system – a change in one level changes the whole. If we focus on one level, namely the chemical, we see that underlying the neural processes of the brain are chemical flows and these flows have a *velocity*. There is, in other words, an underlying set of chemical velocities. These velocities can be increased or decreased. A catalyst, by orienting the appropriate nuclear bonds, can enable a chemical process to move more quickly, requiring less energy. What would happen to our vision were we to introduce the appropriate catalysts or set of catalysts such that all chemical velocities are increased?

Before we introduce our catalysts into the brain, let us take a cube that is rotating slowly, and gradually increase the speed of rotation. The cube will transition through a series of forms with increasing numbers of "serrated" edges as in Figure 2.7. We have started from a figure of four-fold symmetry (period = 4), and transitioned through figures of 4n-fold symmetry – 8, 12, 16, etc., (as n increases from 2, 3, 4…). At a high enough rotation speed, we eventually perceive a *cylinder* surrounded by a fuzzy haze (a cylinder/circle is a figure of infinite symmetry). With our cube in this spinning, fuzzy cylinder form, let us now gradually increase the chemical velocities of the brain via our catalysts. We would expect to see the spinning cube transition in the reverse order of Figure 2.7, through figures of (decreasing) 4n-fold symmetry (24, 20, 16, 12, 8) until, assuming a certain strength of catalyst, we are again perceiving a cube rotating slowly. Were we simultaneously watching a "buzzing" fly, its wings would gradually slow until, perhaps, it is perceived as "heron-like," slowly flapping his wings We have changed the time-scale of our perceptual world. We have also gradually changed the "quality" of both the cube and the fly.

We have now entered a realm analogous to the changing "space-time partitions" of Special Relativity. In relativity, as one observer (perhaps in a rocket) moves at increasingly higher velocity relative to another who is stationary, the rocket-observer's distance and time units change. That is, his space-time partition changes. It is only invariance laws that hold across these partitions. With an adjustment of space units and time units, the law d=vt (distance = velocity x time) holds in the space-time partitions of both observers. Invariance laws are the only realities of Einstein's universe. The invariance law, in our spinning cube case, is a figure of 4n-fold symmetry in every perceptual space-time partition as the brain's chemical velocities change. This is the deeper significance of Gibson's insistence on invariance laws. The perceptual "space-time partition," as we have just seen, is also capable of being variable due to changes in *chemical* velocities. If it is possible in principle, *we must assume nature has allowed for it*, and in so doing, nature must rely on invariance laws to specify events. The frog watching a barely moving fly the frog is about to zap effortlessly with his tongue, the chameleon about to do the same, the chipmunk scurrying by, each with much higher metabolisms – all live in quite different space-time partitions from ours, but all perceive by the same laws of invariance.

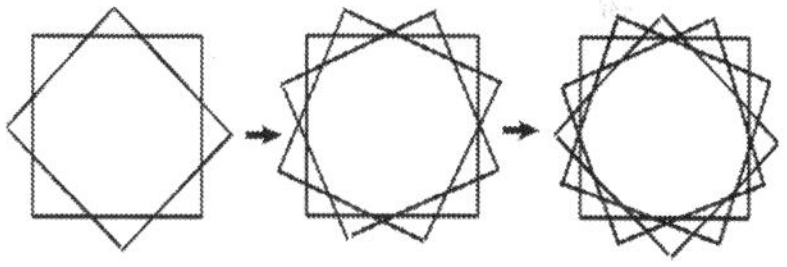

Figure 2.7. Successive transformations of the rotating cube (2-D view) through figures of 4n-fold symmetry as angular velocity increases.

Now let us do another thought experiment on the relativistic implications of all this. Imagine two observers, Mr. A and Mr. B, sitting side by side. Both are watching a buzzing fly. In Mr. B, however, we have introduced our catalysts, increasing the chemical velocity of his neural processes. For both A and B, the razor's edge of the "present" instant of the dynamically changing matter-field we call the universe is the same. Mr. A, in our normal space-time partition, sees the usual "buzzing" fly, where two hundred or so wing-beats/second have been resolved into a blur. Mr. B, with process velocity increased, sees a fly barely flapping his wings. What Mr. A sees as "present" – hundreds of wing beats collapsed as a blur – is vastly in the past for Mr. B who may be focused on the very last wing beat of the whole set comprising A's "blur." Have all these preceding wing-flaps ceased to exist for B? Then why not for A?

Finally, let us take this transformation on Mr. B, wherein we are raising his velocity of processes (or his "energy state"), to its logical endpoint. We left Mr. B with the fly barely flapping his wings. Continue raising B's energy state. (We shall ignore limitations of physics that occur eventually at some point.) The fly becomes immobile, then is perceived in its liquid crystalline, vibrating structure, then becomes an ensemble of electrons whirling, then spinning quarks, etc. Finally we arrive at what I would term the "null scale" – the smallest imaginable scale of time. This scale looks nothing like our normal world. At this null scale, Mr. B and the fly are simply "phases" in the same, vast, dynamically changing field. We shall need to briefly return to the null scale later.

The Relativity of Virtual Action

In Bergson's succinct phrase, perception is *virtual action*. So he would begin his argument in *Matter and Memory*:

> [Objects] send back, then, to my body, as would a mirror, their eventual influence; they take rank in an order corresponding to the growing or decreasing powers of my body. The objects which surround my body reflect its possible action upon them. [5]

One must reflect a moment on the meaning of this statement, gaze around the room, note the myriad of objects arranged at various distances from the body. The increasing distance of some from the body, if nothing else, represents an increasing time just to move to reach them. As well, various objects represent spatial manipulations by the body with each, from forks, to lamps, to books, to tables, etc. The possible action provides a form of *selection mechanism* relative to the vast field of matter in which the body is embedded. We must recall that Bergson visualized the universal field of matter as a vast field of "real actions." Any given "object" acts upon all other objects in the field, and is in turn acted upon by all other objects. It is in fact obliged to be:

> ...merely the road by which pass, in every direction the modifications, or what can be termed *real actions* propagated throughout the immensity of the entire universe.[6]

As Bergson argued, for all organisms in this field and their perception, "...the real action passes through, the virtual action remains."[7] If the brain is a reconstructive wave passing through this holographic matter-field, there must be this principle of selection from the vast amount of information – the real actions – in the "hologram," ultimately for the *modulation* of this reconstructive wave.

We saw earlier that the visual areas of the brain, V1 thru V5, are interconnected, modulating each other's processing ("re-entrant") and themselves modulated by connections from the motor areas. Indeed a large number of recent findings have reinforced the general concept that the objects and events of the perceived world are in a real sense mirrors of the biologic action capabilities of the body, while the appreciation of the importance to visual computation of re-entrant connections from motor areas to visual areas has also grown.[8] But the implications are more radical than currently understood.

There is a ratio, known as *tau*, that exists over flow fields. For the record only, it is defined by taking the ratio of the surface (or angular projection) of the field at the retina, r(t), to its velocity of expansion at the retina, v(t), and its time derivative. For a bird coming in for a landing, this value specifies the severity of impact, and the bird can use it to modify his flight and create a soft landing. For a pilot, it is essential in landing a plane (Figure 2.8). As the coffee cup is moved over the table towards us, this tau value specifies time to contact and provides information for modulating the hand to grasp the cup. Tau is an example of the "information in the light" that must be utilized by the action systems of the brain to specify possible (virtual) action.

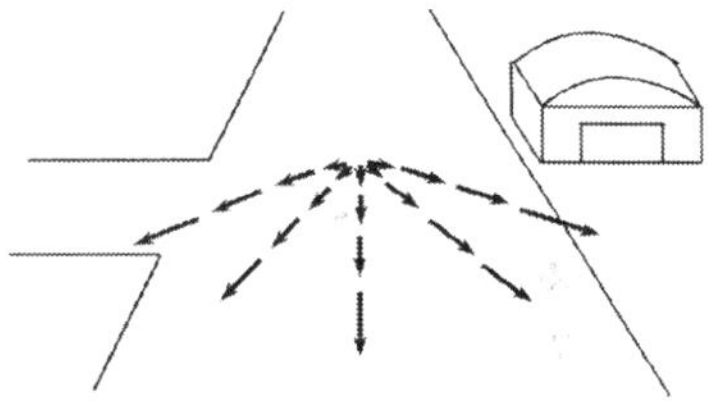

Figure 2.8. The Tau value existing over a flow field is used by a pilot to guide in the plane.

Earlier we considered the effects of introducing a catalyst into the chemical makeup of the body/brain dynamics, with effects on the perceptual "space-time partition." In the possibility of the fly becoming a heron-like fly, slowly flapping its wings, we already previewed the relativistic aspect of this principle. The fly is always presented to us at a certain scale of time. Let us complete the implications, for the time-scaling of the external image is not a merely subjective phenomenon – it is objective, and has objective consequences realizable in action.

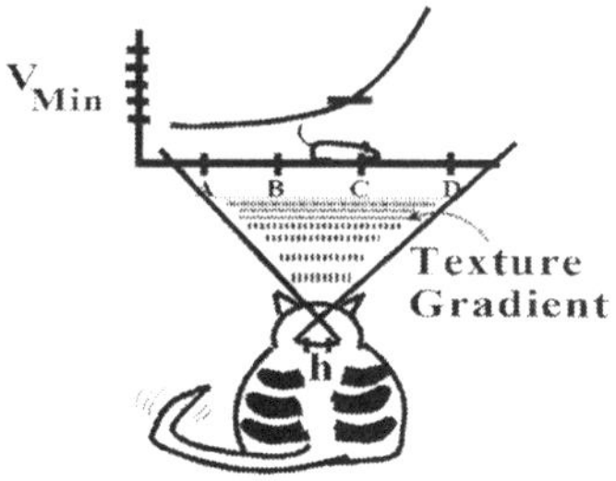

Figure 2.9. Hypothetical function describing the minimum velocity required for the cat to intercept the mouse at D.

Let us picture a cat viewing a mouse traveling across the cat's visual field (Figure 2.9). As we have seen, there is a complex, Gibsonian structure projected upon the cat's retina. There is the texture density gradient over the surface between mouse and cat. The size constancy of the mouse as it moves is specified by the constant ratio of texture units it occludes, while the tau value is specifying the impending time to contact with the mouse, with its critical role in controlling action. All this and much more is implicitly defined in the brain's "resonating" visual and motor areas. How is the information related to action we must first ask?

Michael Turvey, a great theorist of perception and motor action, described a "mass-spring" model of the action systems. For example, reaching an arm out for the fly is conceived as the release of an oscillatory spring with a weight at one end. "Stiffness" and "damping" parameters specify the end-point and velocity of such a spring system. These constitute "tuning" parameters for the action systems of the body. The tuning parameters "bias" the hand-arm to release at certain velocity and stop, just as a coiled spring.[9]

The track (ABCD) along which the mouse runs projects upon the cat's retina (the little "h" in Figure 2.9). For computing the velocity of the mouse, we need: velocity = distance/time. Here the distance (d) being traversed on the retina is equivalent to h (the track/line projected on the retina). But if we wish to use h for distance in our equation to obtain the velocity of the mouse, we have an ambiguity. Similar to the problem with Berkeley's line ABCD in Figure 1.8 (Chapter I), we can move this horizontal track, along which the mouse runs, even closer to the cat, yet the horizontal projection (h) on the cat's retina is exactly the same. Any number of such mice/tracks at various distances project similarly to h; h is always the same. To solve for the velocity of the mouse, the needed muscle-spring parameters must be realized *directly* in the cat's muscular structures via properties of the optic array, e.g., the texture density gradient across which the mouse moves and the quantity of texture units he occludes.

At our normal scale of time, we can envision a function or curve (in Figure 2.9) relating the minimum velocity of leap (V_{min}) required for the cat to leap and intercept the mouse at D as the mouse moves along his path. The closer the mouse gets to D, the faster the cat must leap, until the minimum velocity is so high, it becomes impossible. But to compute the velocity of the mouse, the body needs one other critical thing. It needs a standard of time.

A physicist requires some time-standard to measure velocity. He could use a single rotation of a nearby rotating disk to define a "second." But if an evil lab assistant were to surreptitiously double the rotation rate of this disk, the physicist's measures of some object's velocity would be halved, for example, from 2 ft/sec. to 1 ft/sec. The body must use an internal reference system. Similar to the rotational velocity of the disk, this internal system must be an internal chemical velocity of the body. Such a system is equally subject to such "surreptitious" changes. We noted that these velocities can be changed by introducing a catalyst (or catalysts) – an operation that can be termed, in shorthand, modulating the body's *energy state*. If I raise the energy state, the function specifying the value of V_{min} for the cat must change. There is a new (lower) V_{min} or minimum velocity of leap defined along every point of the object's trajectory, and therefore the object, *if perception is to display our possibility of action with ecological validity*, must appear to be moving more slowly. The mouse is perceived as moving more slowly precisely because it reflects the new possibility of action. If this were not the case, we would be subject to strange anomalies. The cat would leap leisurely; the mouse would long be gone. The cat's perception would not be ecologically valid; it would not specify the appropriate action available.

If the fly is now flapping its wings slowly, the perception is a specification of the action now available, e.g., in reaching and grasping the fly perhaps by the wing-tip. In the case of the rapidly rotating cylinder with serrated edges (once a cube), if by raising the energy state sufficiently we cause a perception of a *cube* in slow rotation, it is now a new specification of the possibility of *action*, e.g., of how the hand might be modulated to grasp edges and corners rather than a smooth cylinder.

Again, the requirement comes to the fore for invariance laws holding across these space-time partitions of perception, in the rotating cube's case, the invariant figure of 4n-fold symmetry in any partition. The hand's modulation is guided by the same invariance information across partitions. As another illustration of this point, we can consider the aging of the human facial profile. This is a very slow event in our normal partition (Figure 2.10). As Pittenger and

Shaw showed, the aging transformation is a *strain* transformation upon a cardioid (a heart shaped figure). We could imagine this aging event as greatly sped up, perhaps as though for a Mr. C who dwells in a much lower energy state where lifetimes pass very quickly. For Mr. C, watching another human, the head is transforming very quickly. Yet, for Mr. C, the aging event is specified by exactly the same invariance law, and his action systems will use the same information should he reach out to grasp the rapidly transforming head.

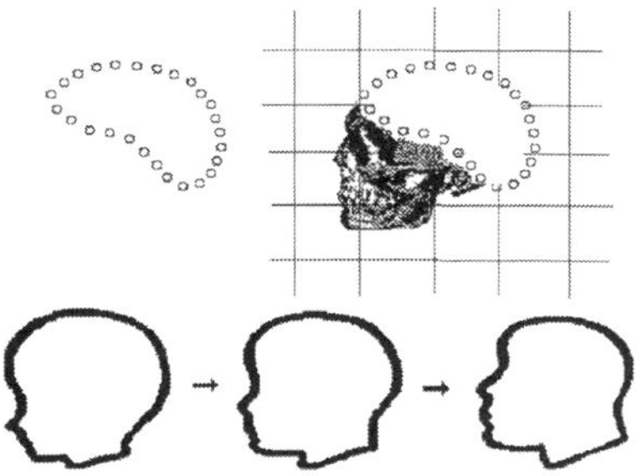

Figure 2.10. Aging of the facial profile. A cardioid is fitted to the skull and a strain transformation is applied. (Strain is equivalent to the stretching of the meshes of a coordinate system in all directions.) Shown are a few in the sequence of profiles generated.

That perception is the display of possible action, that a heron-like fly specifies the possibility of grasping it by the wing-tip – this is a testable implication of the principle of perception as virtual action. The theory described here naturally generates it. I can think of no others that do, though in retrospect the implication will seem obvious. For the computer model of mind, it would merely be an ad hoc afterthought or theoretic "epicycle," not that the computer model can account for the image of the external world in the first place, let alone its scale of time. This is not to say that our current understanding of the "catalysts" at work is anywhere close or will be easy, but the prediction is testable in principle nonetheless.[10]

Time-Extended Events

The specification of the scale of events and the form of events are fundamental aspects of quality (or qualia) in conscious perception. The "buzzing" fly is one quality; the heron-like fly another. The wobbly not-cube is one quality, the rigid cube-in-rotation, another. Were the cube the color "red," Mr. B would be seeing a more "vibrant" red, as his vision has moved closer to the (400 billion/second) oscillations and developing waves of the electromagnetic field supporting this red color. Philosophers of consciousness ignore all this in their search for the explanation of the "qualia" of vision. In truth, the examples we have been discussing here on form reduce certain philosophical conceptions of qualia, where, for example, qualia are virtually akin to little atoms residing in certain neural structures waiting to be configured to look like the external world, to an absurdity.[11] One wonders what "form qualia" could be, other than the forms themselves? Are there "qualia" for every possible scale of time? For wobbly cubes? For spinning cubes at each possible rate? But there is something yet more fundamental being ignored. We must explain how the brain specifies *time-extended* events – "rotating" cubes, "buzzing" flies, or the "singing" notes of violins. This is to say we meet the question of how perception can be of the *past*.

A favorite answer is that the brain takes samples of the cube as it rotates and stores these in its memory. There are many problems with this. If the brain stores a series of snapshots of the rotating cube, it has stored a set of *immobile* pictures, as though we have laid a series of snapshots out, in order, on a desktop, just sitting there. How do we now account for the motion? By some form of internal scanner scanning across the snapshots? Then we have yet another infinite regress. We would now need to explain how the scanner perceives motion.

Another problem is presented by the strobed, rotating cube. The cube is seen as indeed a cube only when the strobe is in phase with (or at an integral multiple of) its symmetry period. A sampler would need to be pre-adjusted to the cube's symmetry period, else we should always see a wobbly cube. This would require ESP or pre-cognition on the sampler's part. And what if there were two or three cubes rotating at different rates?

Discarding sampling as a possibility, another explanation is simply to assume a natural "time-extension" of the neural processes. Somehow, the neural processes computing the form of the rotating cube just "exist" or "last" long enough – are unbroken and continuous enough – for us to see a "rotating" cube, not just a series of instantaneous snapshots, where each snapshot exists only by itself, the previous snapshot having just disappeared. Let's look at this, vague a process as it is. The brain is part of the ever transforming, ever changing matter-field. What is the time-extent of the "matter" of this field? Is it one nanosecond? One second? One minute? One year? One century? The fact is, we think the brain must *store* our experience precisely because the matter-field has *no* extent backwards into the past. Time is seen as a series of instants. Each present instant falls away instantly into non-existence (into the past) as the next arrives. The past is, by this definition, *the non-existent.* The brain then only exists for an instant, an instant in which it must store the present. How long is this instant? Let us consider where this concept of the "series of instants" comes from.

Abstract Space and Time

The concept of the "instant" lies in an *abstract* space and time. Bergson's vision of the world, before the abstraction, is deeply concrete and vastly connected:

> Matter thus resolves into numberless vibrations, all linked together in uninterrupted continuity, all bound up with each other, and traveling in every direction like shivers through an immense body. In short, first try to connect together the discontinuous objects of daily experience; then, resolve the motionless continuity of their qualities into vibrations on the spot; finally fix your attention on these movements, by abstracting from the divisible space that underlies them and considering only their mobility... you will thus obtain a vision of matter...[12]

Abstract space, he argued, is derived from the world of separate "objects" gradually identified, ironically enough, by our perception. It is an elementary process, for perception must partition the continuous, dynamic field which surrounds the body into objects upon which the body can *act* – to throw a "rock," to hoist a "bottle of beer," to grasp a "cube" which is "rotating." This fundamental perceptual partition into "objects" and "motions" – at a particular scale of time we should note – is reified and extended in thought. The separate "objects" in the field are refined to the notion of the *continuum of points or positions*. One can visualize this continuum as your desktop, where the desktop is comprised of row after row of little points, from one end to the other, much like the gradient picture of Figure 1.7. As an object moves across this continuum, as for example, my hand moving across the desk from point A to point B, it is conceived to describe a *trajectory* – a line – consisting of the points or positions the hand traverses. Each point momentarily occupied is conceived to correspond to an "instant" of time. Thus arises the notion of abstract time – the series of instants – itself simply another dimension of the abstract space.

This space, argued Bergson, is in essence a "principle of infinite divisibility." Having convinced ourselves that this motion is adequately described by the line/trajectory the object traversed, we can break up the line (space) into as many points as we please. But the concept of motion this implies is inherently an infinite regress. To account for the motion, we must – between each pair of static points/positions (i.e., *immobilities*) supposedly occupied by the object – re-introduce the motion, hence a new (smaller) trajectory of static points – ad infinitum.

The instant, as simply another form of space, is infinitely divisible. By the logic of abstract space and time, the very logic which motivates our theoretical need to *store* past instants in the brain in the first place, the instant over which the brain has a time-extent is divisible unto the smallest period of time possible, ultimately to a mathematical point. This mathematical point is the actual extent in time of the neural processes of the brain, or equivalently, the actual time-extent of the brain as matter. The neural processes of the brain can only be static instant after instant after static instant, each with the extent of a mathematical point, each disappearing as the next arrives. There is no way – to be consistent with the logic which drives us to store memory in the brain in the first place – that neural processes are continuous enough to support the perception of a "rotating" cube. Yet there is no easily usable notion of a memory, brain storage, "iconic" storage, or snapshot storage holding area, or whatever, that works for explaining our ongoing vision of the rotating cube.

Motion as Indivisible – or Primary Memory

Motion, Bergson argued, must be treated as *indivisible.* He noted that the ancient Greek philosopher, Zeno of Elea, had created a set of four paradoxes based on abstract space and time, precisely to stick the noses of his contemporaries into the problem. There is a mythology that modern philosophy and/or mathematics has done away with this problem, Bertrand Russell being one example. Russell, undoubtedly to the shock of Zeno, actually accepted the fourth paradox (where a motion is declared "the double of itself") as an actual, physical reality, though in fact it has the same root cause as the other three, and changed his position about the rest at various times. The paradoxes are still very much operative beneath modern thought.

A famous one is Achilles and the hare. Chasing the hare, with each step Achilles covers half the distance between himself and the hare. The halved-distances get smaller and smaller, but no matter how small the half, there is always a bit left. He never catches the hare. In the "paradox of the arrow," Zeno noted that the arrow, at each instant of time in its flight, corresponds with a stationary point in space. As such, argued Zeno, the arrow "never moves." These paradoxes of Zeno, Bergson held, had their origin in the logical implications of an abstract space and time; they were Zeno's attempts to force recognition of the invalidity of this treatment. When Achilles cannot catch the hare, it is because we view his indivisible steps through the lens of the abstract trajectory, the line or space each step covers. We think of the abstract space traversed. We then propose that each such distance can be successively halved – infinitely divided in other words. In the paradox, Achilles never reaches the hare. But in the concrete world, Achilles moves in an indivisible motion; he most definitely catches the hare.

The abstract conception of space and time, however, is even further rarified. The motions are now treated as *relative*, for we can move the object across the continuum, or the continuum (or the coordinate system) beneath the object. Motion now becomes *immobility* dependent purely on perspective. All *real*, concrete motion of the matter-field is now lost. But, Bergson argued, there must be *real* motion. The universe, the entire matter-field, must dynamically change and evolve over time. Trees grow. Flowers bloom. People get older. Mountain ranges appear. Stars shrivel and die. He would insist then:

> Though we are free to attribute rest or motion to any material point taken by itself, it is nonetheless true that the aspect of the material universe changes, that the internal configuration of every real system varies, and that here we have no longer the choice between mobility and rest. Movement, whatever its inner nature, becomes an indisputable reality. We may not be able to say what parts of the whole are in motion, motion there is in the whole nonetheless.[13]

We must, he argued, view the entire matter-field as a *global* motion over time. We must see the *whole* changing, he argued, "as though it were a kaleidoscope." We want to ask if individual object X is at rest, while individual object Y is in motion. But both "objects" are simply arbitrary partitions, phases in this globally transforming field. As such, the "motions" of "objects" are seen as *changes or transferences of state* – rippling waves – within the dynamic motion of the whole.

The motion of this whole, this "kaleidoscope" as Bergson called it, cannot be treated as a series of discrete states or "instants." Rather, Bergson would argue, this motion is better treated in terms of a melody, the "notes" of which permeate and interpenetrate each other, the current "note" being a reflection of the previous notes of the series, all forming an organic continuity, a "succession without distinction," a motion which is indivisible.

From this perspective, there is a "primary memory." It is a property of the matter-field itself and of its melodic motion. This primary memory underlies the motion of the rotating cube, even the motion or flow in the neurons of the brain. The motion of the field, of which the rotating cube is just a phase, does not consist of discrete instants that fall away into the past, or into non-existence. For this reason, the brain, with its reconstructive wave, is perfectly able to specify a past transformation or motion of the matter-field. The wave can specify "rotating" cubes or the "singing" notes of violins. We are always viewing the past.

Physics and the Abstraction

Abstract space/time is a "projection frame" for our thought, derived from the necessity for practical action. The theoretical world of physics became a population of tiny "objects" moving from "point" to "point" in the abstract space, "instant" by "instant" in the abstract time. This is *classical* space and time. It is also, unfortunately, still the time of relativity theory. For physics, the effort to break from this projection frame has been very real. To Bergson, "*...a theory of matter is an attempt to find the reality hidden beneath... customary images which are entirely relative to our needs...*"[14] For physics, the "customary images" of the abstraction have been the ultimate barrier.

What is a "particle," Bergson asked, but the extension in thought of this bodily perceptual process by which useful "objects," be it spoons, coffee cups or bottles of beer were first identified in the whole. It is a concept derived purely for practical action which will never, imported into the realm of pure knowledge, explain the properties of matter.

> But the materiality of the atom dissolves more and more under the eyes of the physicist. We have no reason for instance, for representing the atom to ourselves as a solid, rather than as a liquid or gaseous, nor for picturing the reciprocal action of atoms by shocks rather than in any other way. Why do you think of a solid atom, and why of shocks? Because solids, being the bodies on which we clearly have the most hold, are those which interest us most in our relations with the external world... [15]

Thus, from a population of theoretical particles – muons, gluons, leptons – physics eventually moved to "quarks." But the quarks, with their various "spin" states, became ever less material, and we are asked to abstract from the spin state of a quark all mass, leaving an abstract mathematical point with its value of "spin." Below this now are postulated the "strings," inconceivably small violin strings as it were, whose harmonics give rise to the whole of the field of matter. And thus, in the end, contemplating the dynamic movement of this field:

> ...they show us, pervading concrete extensity, modifications, perturbations, changes of tension or of energy, and nothing else.[16]

Simultaneously, the concept of a *trajectory* of a moving object also faded. This no longer exists in quantum mechanics. One can determine through a series of measurements only a series of instantaneous positions, while simultaneously renouncing all grasp of the object's state of motion, i.e., Heisenberg's famous principle of uncertainty. As de Broglie would note, writing his comparison of Bergson to current concepts of physics, the measurement is attempting to project the motion to a point in our abstract continuum of points or positions, but in doing so, we have lost the motion. Thus Bergson noted, over forty years before Heisenberg, "In space, there are only parts of space and at whatever point one considers the moving object, one will obtain only a position."[17]

The physicist, Laurent Nottale (*Chaos and Solitons*), simply notes a proof (by Richard Feynman) that the motion of a particle is continuous but not differentiable.[18] Hence, he argues, we should reject the long held notion that space-time is differentiable. He opts for a fractal approach – indivisible elements which build patterns. The essence of differentiation is to divide (say, a motion from A to B, or the slope of a triangle) into small parts. This operation is carried out with smaller and smaller parts or divisions. It is understood that the divisions can be infinite in number, infinitely small. When the parts or divisions have become so minute, we envision "taking the limit" of the operation – obtaining the measure of say, "instantaneous" velocity, or slope, etc. To speak of non-differentiability is to say "non-infinite divisibility." We have something – indivisible. To state that space-time is non-differentiable another way, we may say the global evolution of the matter-field over time is seen as non-differentiable; it cannot be treated as an infinitely divisible series of states.

Lynds (*Foundations of Physics Letters*) now argues, echoing Bergson, that there is no precise static instant in time underlying a dynamical physical process.[19] If there were such, motion, changes in the magnitude of forces, momenta, etc, would not be possible, as all would be frozen static at that precise instant, and remain that way. In effect, such an instant would imply a momentarily static universe. Such a universe (shades of Zeno's arrow that does not move) is incapable of change, for the universe itself could not change to assume another static instant. Consequently, at no time is the position of a body (or edge, vertex, feature, etc.) or a physical magnitude precisely determined in an interval, no matter how small, as at no time is it not constantly changing and undetermined. It is by this very fact – that there is not a precise static instant of time underlying a dynamical physical process or motion – that variation in magnitudes is possible; it is a necessary tradeoff – precisely determined values for continuity through time. It is only the human observer, Lynds notes, who imposes a precise instant in time upon a physical process. Thus, there is no equation of physics, no wave equation, no equation of motion, no matter how complex, that is not subject to this indeterminacy.

With this view, there can be no static form at any instant, precisely because this static instant does not exist. The brain cannot base its computations on something that, to it, does not exist. The brain is equally embedded in the transforming matter-field, i.e., it is equally a part of this indeterminacy. It can only be responding to invariance over a non-differentiable, indivisible flow or change. This non-differentiable, indivisible flow, with its "primary memory," allows the specification of events with a time-extension receding into the past.

The Classic Metaphysic

Abstract space and abstract time form what can be termed the "classic metaphysic." It is this metaphysic that resides behind the entire discussion of qualia and the "hard problem" of Chalmers. As noted, the end result of this "principle of infinite division" that abstract space represents, even could we legitimately conceive of an end of such an operation, ignoring the mathematical hand waving of taking a "limit," would be at best a mathematical point. At such a point, there could exist no motion, no evolution in time of the field. Further, as every spatially extended "object" is subject to this infinite decomposition throughout the continuum, then we end with a completely *homogeneous* field of mathematical points. The continuum of mathematical points then, both spatially and temporally, can have no qualities – qualities at the least imply heterogeneity, i.e., multiple differences.

That this is indeed the framework that the qualia debate participants have tended to work within is attested to by a very common starting point, namely that the matter-field contains no qualities – objects have no color, there are no sounds, etc. This framework is also betrayed by the fact that the vast preponderance of examples of qualia are static – the "redness" of red, the taste of cauliflower, the feel of velvet, the smell of fresh cut grass. Seldom are qualities of *motions* ever discussed, e.g., the "twisting" of leaves, the "gyrations" of a wobbling, rotating cube, the "buzzing" of a fly. Hardcastle describes things with hints of the dynamic: "I go to the symphony… I see the conductor waving her hands, the musicians concentrating, patrons shifting in their seats, and the curtains gently and ever-so-slightly waving."[20] But the hint of time here is never followed, in fact, it is dismissed as a problem for her subsequent analysis. This glaring lack is coordinate with the fact that an abstract "time" that is simply another dimension of the infinitely divisible space – a line of infinitely minute point-instants – is equally completely homogeneous. Any "motion" in this space, logically, has no duration greater than a mathematical point, then another point, then another. In fact, then, the debaters universally fail to realize that the perceived time-extents of these motions – the rotating cube, the buzzing fly, the whirling of the coffee surface with circling spoon – are equally *qualities* that arise, just as problematically as the "static" colors of objects, in the homogenous time dimension of infinitely divisible instants in this continuum.

The origin of the classic metaphysic traces to Galileo. Galileo's crucial step was to suggest that the real world is made only of *quantitative* aspects while other empirical aspects – the qualities of the experienced world – are somehow created by "the living organism." Shape (form) is considered part of the quantitative realm and thus considered part of the "real" – not a quality therefore and not part of the hard problem. Implicit within Galileo's statement is the distinction between *primary* properties and *secondary* properties, the former related to quantity and "real," the latter related to quality and only in the mind. This exact structure runs ubiquitously in the statements of the philosophers of consciousness today.

Form as Qualia

Yet, Galileo, as he initiated the classic metaphysic, was even wrong when he assigned shape or form to his quantitative continuum while thinking he was excluding qualities there from. There is nothing static in the ever-transforming material field. Edges, vertices or surfaces do not exist in an instant. Nor color. Again, there are no "instants." Again, the brain, simply a part of this transforming flux, cannot use in its computations that which for it does not exist. Even form can only be derived by imposing constraints (invariance laws) over ever flowing velocity fields.[21]

While the earlier discussed "coding problem" itself has never been truly understood, with the result that innumerable "solutions" have been generated to the hard problem that are blind to a coding dilemma at their core, the misconception of static form is another drag. Derived from the classic metaphysic and Galileo's mis-assignment of form to the mere "quantitative," it also underlies a fundamental failure in the qualia debate participants. This is the failure to grasp that the issue being addressed is the problem of the origin of the *image* of the external field. The argument here, that the origin of the external image is *the* problem, is incomprehensible to them, hence the significance of Bergson's model of the brain as a reconstructive wave, specifying a time-scaled image of an inherently qualitative field is totally lost upon them. The misconception, despite the fact that the origin of the image itself as a problem was clearly understood by earlier philosophers, is harbored by virtually all in the debate today. Martine Nida-R□melin states it clearly in her essay. She feels forced to differentiate between color as an "appearance property" (and therefore a qualia problem) and shape, which she says is not such a property (therefore not a qualia problem).[22] Though it is unclear to me how any form can fail to have some color qualia at its base, even a skeletal cube composed of stick-like edges in some shade of gray, it appears that all seem to think that the origin of the image of the *forms* (or shapes) of the external world is no problem – these are easily "computable" and hence the image itself is no problem, only its "qualities." They fail to grasp that the origin of the image of the forms in the field or of the objects in the field is just as much a problem as the (other) "qualities" of the field – the "rednesses," the "velvets," etc., etc. *None* of these is simply computable. At the null scale of time, the material field, in its massive, continuous dynamic flux, looks nothing like the *image* we have of it at normal scale. The "buzzing" fly of our scale is simply a mass of shivering field oscillations. Technically, the field, at its null or "natural scale," is non-image-able. *It is the origin of our image of this field, any image, that is the problem.*

The brain is integrally a part of the abstract continuum of the classic metaphysic. Therefore, when light rays strike objects termed eyes in the brain, the abstract, homogeneous motions of the external matter-field, all reducible in time-extent to mathematical points, simply continue in the portion of the field called the "brain." Nowhere in the brain, taken as part of the abstract continuum, can there be anything but more homogeneous points/instants. As we have seen, there can be no actual time-extent of motions through the nerves, no "continuity of time-extended neural processes" – the logical time-extent of any neural process, as noted, is never more than a mathematical point, then another, then another. However one views these motions within the brain, e.g., as maintaining some structural correspondence or isomorphism relative to the always past transformations in the external field or as the processing of invariants in this structure of field motions relative to the body's action systems, it changes nothing. Within the brain, taken as a part of this abstract, homogenous continuum, we can never derive qualities, whether qualities of objects (colors, smells) or of time-extended motions (ignoring that the "object" *is* a motion). We cannot explain how we see a cube "rotating" let alone a "blue" cube. Therefore, all qualia are logically forced, within this metaphysic, into the non-physical, or the mental, or somewhere, anywhere but the abstract continuum. But the step by which this generation of events unto and into another realm can occur, *within the confines of the metaphysic*, remains a dilemma. The structure of the metaphysic makes the step impossible, while leaving the nature of realms outside the structure – e.g., the "mental" – forever incapable of definition or of use to the science that currently operates precisely within this metaphysic.

The alternative, the *temporal* metaphysic, resolves all of this. The time-flow of the matter-field – indivisible, non-differentiable – is inherently qualitative. Objects are but flows in this field. The brain, at a certain energy state, as a reconstructive wave, is specifying a past motion of this qualitative field – at a scale of time – reflective of the possibility of the body's action at this scale.

Illusions and Direct Realism

I am compelled here, before we go on, now that the concepts we need are in place, to devote a bit more discussion to the illusions of indirect realism. The subject of illusions is the great sticking point for the holders that "all is generated by the brain." If direct realism is construed as simply seeing "what is there" in the matter-field, this view is untenable in the face of multiple lines of evidence. As we have seen, given the inherent uncertainties of information with which it deals, the brain is in fact computing an optimal percept. On this basis alone, the brain's specification of events in the matter-field is based on probabilities. But to construe direct realism as implying that we simply see "what is there" is in fact simply an expression of what is termed *naïve* realism. Given the inherent uncertainty of measurement, nothing is simply "there." Further, as we have just seen, it is always a specification of the past, therefore, already a memory.

This aspect – that of perception already being a memory – requires me to anticipate the next chapter slightly. If the brain supports a reconstructive wave in a holographic field, then this wave is also capable of reconstructing past events in this field. The invariance laws or invariance structure of a "present" event is simultaneously creating a wave that is reconstructive of past events with similar structure. Bergson thus argued that perception is always permeated with memory experience (he saw the "flow of memories" to perception as a "circuit"). The initially indistinct words to a song, with a bit of a clue, from then on are perceived as "perfectly clear" whenever we subsequently hear the song. The "bear" near the trail in the woods, seen to be a tree stump on closer approach, would also be a reconstructed experience, now fused in the perception. This is already a form of "filling in," a phenomenon considered very problematic for direct realism. Spatial filling in can occur when a blind spot in the visual field (perhaps due to neural damage, say, a scomata) is not perceived by the subject as an empty area because the brain fills it in by completing the surrounding pattern.[23] A black line containing, in actuality, a small gap, might be seen as complete. The natural question, as asked by the philosopher, John Smythies, is how is it possible to say that the area (e.g., the line) on either side of the blind spot is seen directly, but the gap (now seen as part of the line) is constructed (or seen indirectly)?[24]

The answer is "quite easily." It can easily be both. We may call this "constructive directness" if one likes. The perception is always a specification of some past state of the field. A reconstructive wave passing through a hologram can specify simultaneously in superposition, wave fronts recorded at Time 1 and Time 2. What is specified is always dependent upon a modulation pattern. This pattern may use or embody some probabilistic constraints or rules, but it is always a specification of a past form of the holographic field. It can easily include material from memory – it *always* does. This material – events of the past coalescing as images – is no more generated by the brain than the original perceptions.

Yarrow et al. experimented with viewing a silently ticking clock, where, during saccades (rapid eye movements), the second hand appears to take longer than normal to move to its next position (as though the hand briefly stopped).[25] Under such conditions, objects presented during a saccade are actually invisible. The visual system appears to be shut down for an instant, but the brain computes what we would have seen during the saccade. Smythies notes that it would be most implausible to suggest, per direct realism, that we see directly only when our eyes are not in saccadic movement. But the answer is that the perception is as direct as ever. During the clock hand's motion relative to a receptive eye, information from the field is taken in, the optimal percept computed, the reconstructive wave/specification is still to the past. During the saccade, the reconstructive wave does not cease – it continues to specify a state of the field based on the information available and the probabilistic algorithm employed by the neural architecture.

Kevin O'Regan is similar in this respect.[26] He noted that an entire page of surrounding text can be changed during a saccade without notice while someone is reading as long as the 17-18 character window the eye is focused upon is undisturbed. He opted to conceive of the environment as an "external memory store" to explain the persistence of the perceived world during saccades. However, the scale of time of this "external" store would be very problematic – would an external fly look like the buzzing fly of our scale, or be flapping his wings like a heron or be a mass of electrons? We can better say that the reconstructive wave and/or the pattern supporting it within the brain is not affected by a substitution of the surrounding text during a saccade with its minute information gathering capacity (44 bits?), the brain's specification yet being to the same states of the past.[27]

O'Regan's observation was extended under the heading of "change blindness."[28] Experiments demonstrated this dramatically. Under the right conditions, an actor in a gorilla suit could walk across a tennis court in the middle of a tennis match which subjects were observing on video, and yet be utterly invisible. Subjects were astounded when the gorilla was pointed out. This led to speculation that our perceived world, in total, is an "illusion," again, simply generated. But that the brain chooses not to employ certain information, such that the specifying reconstructive wave contains nothing about the field relative to the gorilla, is insufficient reason to call into the question the directness of perception. There is, in fact, *always* a ton of information in the holographic matter-field – the sea of real actions – that is not being used, and not part of the virual action being specified.

I should note here how similar the "invisible gorilla" is to a form of hypnosis. The great hypnosis expert, Cleve Backster, on stage, gave a hypnotic suggestion to a subject picked from the audience such that Cleve would be invisible. To the subject, Cleve was invisible, even to the point of being completely freaked out as he observed a cigarette which Cleve lit up as floating unsupported, with its smoke, in mid-air. Again, this can be looked at as simply a modulated or filtered specification, not something that implies the brain generates all reality.

There are other more problematic phenomena. Kovacs et al. constructed two pictures – picture A showing a group of chimpanzees, B showing jungle foliage. If A is shown to one eye, then B to the other, in alternation, the subject, as expected, sees A (chimps), then B (jungle) in alternation.[29] Two pictures C and D were also constructed, where C contained portions of A and B, and D contained the complimentary (unused by C) portions of A and B. Presented C (an intermingling of jungle/chimps) in one eye and D (the complimentary intermingling) in the other, what the subject now sees is not C, then D, but rather, as before, A (chimps) then B (jungle). This is a clear construction by the brain. That the subject, as Smythies notes, is not actually seeing what is "out there" at any given instant, is true. The brain is nevertheless specifying, via information received, past states of the field. Yes, it is a garbled specification of the field, just as a reconstructive wave modulated to imprecise or non-coherent frequencies can specify composite images of A and B. Yes, the famous Necker cube constantly flips in orientation. The information projected to the eye is inherently ambiguous, mapping to two possible orientations of the cube. The brain's specification, at each "flip," is yet to a state of the field.

It is undeniable, as Smythies argues, that the perceived image of the world is the result of massive neurocomputations. What *is* deniable is the final step of this process proposed by the indirect realist, namely "the generation of the sensations that we experience in consciousness."[30] What is meant here is the "generation of the visual image," or more simply the "*generation* of consciousness." This step does not exist. The entire notion of so-called visual "sensations," as Gibson argued, is a misleading alley. The neurocomputations generate no "sensations." The neurocompuations generate no "images." The neurocomputations generate no "projections" back to the coffee cup as Crooks noted in contemplating Figure 1.1. The indirect realists have *never* considered time and the properties of the time-evolution of the matter-field; they have no answer for, nor even thought about, the memory that supports the rotating cube. The neurocomputations support – are involved in – a complex, very concrete reconstructive wave, a wave specifying a past, time-scaled form of the non-differentiable motion of the field.

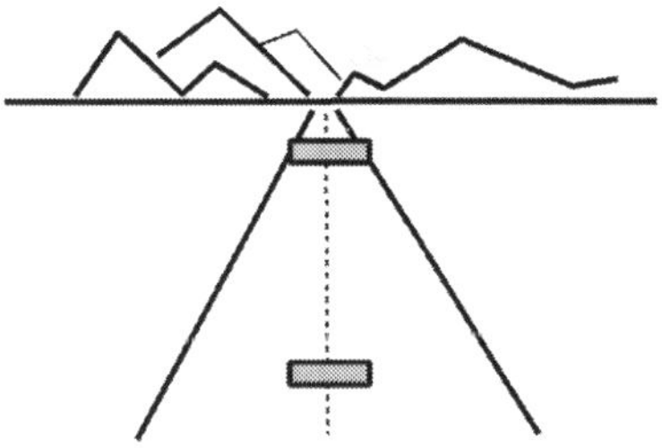

Figure 2.11. The Ponzo "Illusion." The far object on the road looks much larger than the near object, yet both are exactly the same size on the drawing and on the retina. Ecologically, in the real world, the far object would indeed be larger.

This optimal percept, based upon invariance information in the field and probabilistic constraints, is, under ecological conditions, extremely veridical, as indeed it must be for survival. It is not nearly as illusory or unreliable as the indirect realist emphasizes. The golf ball, sent rolling by Tiger Woods across the undulating green and grain, is guided by precise specification of the world. The indirect realist is wont to emphasize the existence of illusions –

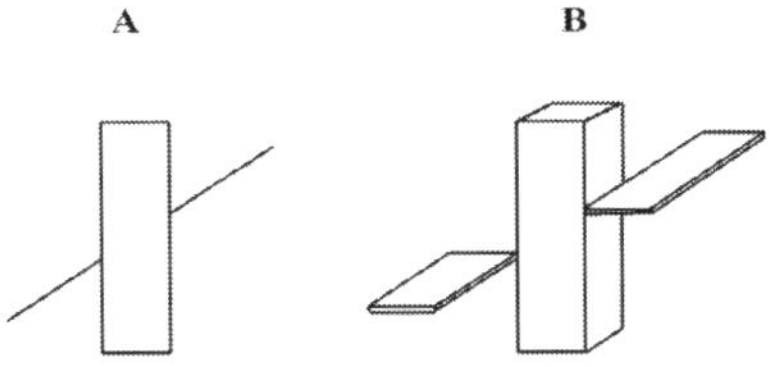

Figure 2.12. Poggendorff Illusion. (A) The oblique lines are actually perfectly aligned – a continuous straight line. (B) The cause of the illusion from a depth computation explanation – the result of the visual system's attempt to interpret A as a three dimensional object, where the oblique line is actually part of two planes, one higher than the other, and thus not in alignment.

the Poggendorff, the Ponzo, etc. – to enhance the constructivist implications for perception. Gibson long argued that these are artifacts, that given an ecological environment, rich with information (invariants), these don't happen, the experience being again "directly specified." In fact, some "illusions" would count as quite valid percepts from a Gibsonian perspective (e.g., the Ponzo, Figure 2.11, where the distant object would occlude far too many texture units to be the same size as the near object). The Poggendorff (Figure 2.12) can equally be interpreted as a specification of a valid state of the (normally) three-dimensional matter-field. Perception theorist, Irvin Rock, opposed Gibson at every turn, arguing for indirect perception.[31] Often his experiments involved information-deprived experimental setups, destroying Gibson's texture gradients, for example, by forcing the observer to judge distances when looking into a darkened room through a peephole (say, the distance of two rods located at different distances on the floor, as the cups in Figure 1.9). He would then argue that inferences or mental operations must be involved. That information-deprived or ambiguous setups exist which demand reliance on probabilistic specifications is certain, in fact, as we have seen, all is probabilistic, however the specification is still to past states or transforms of the matter-field.

Finally, under the out-of-phase strobe, the wobbly, plastic, not-cube is indeed an illusion. Yes, we can ask, how can this yet be an image *in the field*, not generated by the brain? As I noted earlier, record the object waves from several objects A, B, C, D and E on a hologram plate, each with a different frequency of reference wave, f_a through f_e. A reconstructive wave with components f_a, f_c, f_e will be specific to a confused, fuzzy superposition of A, C and E. The information is there, within the hologram; it is a legitimate specification of an "object" via available information. Similarly, even just one second of the strobed, rotating cube is in reality an enormous series of (interpenetrating) states within the external, ever changing 4-d holographic field. Each "state" in this vast, time-extended series contains a slightly different orientation of the cube. Under the out-of-phase strobe, the optimal specification is to something like a superposition of the states of this series (at a scale of time) wherein the flat sides, rigid edges and sharp vertices are lost – the wobbly, plastic-like not-cube. Again, the brain is not projecting or generating an image. Even illusions – the great redoubt of indirect realism and for the ubiquitous conception that the brain must be generating the image – are but optimal specifications of events in the field, where the "image" is precisely where it says it is – in the external field.

The Color of the World

The widely received view that the objects of the physical world are not actually colored obviously has roots in the classic metaphysic. The (white) coffee cup resting before us possesses no color. The (brown) coffee possesses no color. Colors only exist as *subjective* (or "secondary") properties of the matter-field. In this view, color experience is in effect a vast illusion for us all. Objects are *represented* as colored in the dark, quiet brain, yet there are no properties of the physical world that the representations are reflecting. No philosopher knows how the "representations" in the dark, quiet brain become colored. If the color is not in the physical world, then the color of the world becomes entirely a mystical process, beyond the realm of science, in a sad realm ruled only by the philosophers, something perhaps to their liking. Of course, these same statements should be equally said of *form*: Objects possess no form, for form too is not a primary property; it too must be only be subjective. As per the wobbly cube, objects are represented as having a form, yet there are no simple properties of objects that this representation of form is reflecting.

The statements on form move us nearer to the absurdity of this entire "no color" conception. If form too is just as much an illusion as color, just a "representation" in the brain, then is form too just subjective, just, "in the mind?" Then what is not "just in the mind?" This is the ultimate absurdity of "indirect realism." This is just idealism – with frantic hand waving attempting to distract our attention from what it actually is. We owe this conception, as usual, to the ever-present abstract space underlying our theories. A set of abstract, homogeneous "objects," (e.g., particles, electrons, quarks) in equally abstract (and relative) "motions" introduces an impassable gap between these objects/motions and quality (or qualia). The particles have become abstract little billiard balls in abstract "motions" denuded of quality.[32]

At the null scale of time, the matter-field may indeed be *near* the homogeneous state envisioned by classical mechanics in its particles with their abstract motions. But as we impose scale, this changes:

> May we not conceive, for instance, that the irreducibility of two perceived colors is due mainly to the narrow duration into which are contracted the billions of vibrations which they execute in one of our moments? If we could stretch out this duration, that is to say, live it a slower rhythm, should we not, as the rhythm slowed down, see these colors pale and lengthen into successive impressions, still colored no doubt, but nearer and nearer to coincidence with pure vibrations? [33]

I am not going to take us here into the many problems surrounding the information that the brain actually uses to specify colors. Semir Zeki hypothesized that the brain is isolating invariants of spectral reflectance.[34] The spectral reflectance profile of an object is given by specifying the percentage of incident light reflected by that object at each wavelength or over particular bandwidths. But there are multiple apparent problems, all supporting the view that there are no simple properties, such as reflectances, that in and of themselves specify color. Metamers, for example, are stimuli having different spectral reflectance distributions that produce the same experienced color. There is also the complex web of similarity relations among colors. Purple is more similar to blue than green, reds more similar to other shades of red. In this complex, there is an opponent structure: red is opposed to green in the sense that no reddish shade is greenish. So also for yellow and blue. There are unique hues (red, yellow, green, blue), and binary hues (purple, orange, olive, turquoise) – said to be perceptual mixtures of the unique hues. All of this appears as a problem for a qualitative field supporting color, for light waves or surface spectral reflectances do not stand in relations to each other that are unique or binary, opponent or non-opponent, etc. All this is only compounded by the fact that, in our normal environment, *the light conditions are continually changing*, yet somehow the colors of the world remain to us perceptually the same. There is no simple mapping from such physical properties to the subjective color experience. But the wobbly cube shows us there is no simple mapping for form either.

We are presented, everywhere beneath our colored world of experience, with what can only be described as massive, unending, continual flux in the physical world. There is no fixity, nothing fixed for the brain to latch on to, only a flux from which it attempts to derive some semblance of constancy, yes, some *invariance*, though even how this is achieved may vary across various individuals. Color, just as form, is an optimal specification of properties of the external matter-field. This must be so, for the brain is only specifying a time-scaled portion of the matter-field flowing in time, and this field is, inherently, Quality.

Quality as the Knife

Quality, Phaedrus had realized, is "a cleavage term."

> You take your analytic knife, put the point on the term Quality, and just tap, not hard, gently, and the whole world splits, cleaves right in two – hip and square, classic and romantic, technological and humanistic – and the split is clean... Sometimes the best analysts, working with the most

> obvious lines of cleavage, can tap and get nothing but a pile of junk. And yet here was Quality; a tiny, almost unnoticeable fault line; a line of illogic in our concept of the universe; and you tapped it and the whole universe came apart....[35]

Bergson, too, had gently tapped on this fault line. The world split into his "dichotomies:"

- Quality vs. Quantity
- Duration vs. Abstract Time
- Extensity vs. Abstract Space

"Duration" was his term for the indivisible flow of time, or *concrete* time, a time where each "instant" flows into, interpenetrates, merges into the next, like the notes of a melody.

> Might it not be said that, even if these notes succeed one another, yet we perceive them in one another, and that their totality may be compared to a living being whose parts, although distinct, permeate one another just because they are so closely connected? ...We can conceive then of succession without distinction, and think of it as a mutual penetration, an interconnexion and organization of elements, each one of which represents the whole, and cannot be distinguished or isolated from it except by abstract thought. Such is the account of duration that would be given by a being who was ever the same and ever changing, and who had no idea of [abstract] space. [36]

The melodic flow of the matter field is inherently Quality. "Quantity" is simply the stripping of objects of quality, be it particles or apples, such that I can treat them all alike. When I count five apples, I strip each individual apple of its individual qualities; I ignore their subtle differences in color, texture, form, or taste. For the sake of arriving at a sum, I treat each apple alike, as simply a "unit." "Extensity" was simply Bergson's term for the *concrete* space that extends all around us – the floor, the rugs, the fields beyond – none of which consists of truly separate "points" or "positions" or "atoms" or "particles."

This great "cleavage" of the world creates two more "dichotomies:"

- Semantics vs. Syntax
- Intuition vs. Intellect

Syntax rules are the foundation of the computer conception of mind. These are simply rules for the manipulation of abstract objects (symbols), denuded of quality, in an abstract space in an abstract time. These have nothing to do, as we shall see in later chapters, with semantics – the basis for meaning in our world. Because the computer conception of mind sits on the right side of this cleavage point, it cannot deal with the terms on other side. It cannot account for semantics, or intuition. This is because this split has an essential feature: each of the terms on the right is only the *limiting case* of the term on the left. Abstract time, for example, is simply the limiting case of real time – it cannot account for real time or duration, it cannot be made to explain it or substitute for it. The lesser cannot account for the greater. Intuition is concrete mind flowing in concrete, indivisible time. Intellect is its limiting case – mind based on symbols manipulated in an abstract space. It is this brittle, intellectual form of mind, in itself not in touch with the real Quality of the world in its flow, that Phaedrus meant by his term "squareness."

Phaedrus, with his great intuition on Quality, became convinced he was blazing a philosophical trail. He had poured through the philosophical literature for precedents, and found little, the best being the great mathematician. Poincare. Poincare was a contemporary of Bergson and he spoke of the great beauty in certain expressions or equations of mathematics. Even here, in mathematics, Poincare was saying, the intellect itself is, and must be ultimately guided by, the perception of quality. But Phaedrus's lack of awareness of Bergson is only testimony to how thoroughly philosophy had buried Bergson.

All this leads us back to the greatest entry in this set of dichotomies:

- Subject vs. Object.

Subject and Object

We have seen that the matter-field is holographic. We have seen that the dynamic transformation of this holographic field is indivisible, where each "instant" interpenetrates the next, as the notes of a melody, and where each note reflects the entire preceding series of notes. There is a deep implication in this. If the state of each point/event in the field reflects the influences of the whole, in fact the *history* of the whole, then, in effect, each "point" at the null scale of time has an elementary awareness of the whole. Bergson called this "pure perception." It is as though, stretched across the universal matter-field, is a vast, vibrant "web" of awareness. It is a highly *coherent* web; the threadlike "fibers" are taut, a light flick with a finger sends reverberations instantly throughout the whole. This "web," with its basic form of awareness and its fundamental, primary memory as the web transforms over time, carries the elementary attributes of Mind.

Note that I am saying the *null scale* of time. I am not speaking of the scale of time as we know it – the scale of cups, symphonies and buzzing flies. It requires all the time-scaling imposed by the brain's reconstructive wave, as we have earlier discussed, to specify a scale. This is true even for frogs and chipmunks, who, by the way, surely perceive the world at their particular scales of time by this same reconstructive wave process and, by this fact, are also conscious. Note too, we are not ascribing some form of "mentality" to these point-events. We are not looking for "proto-conscious" particles, or tiny "perceiving protons." This is the desperate move of certain current philosophers to explain conscious perception. We only require this elementary awareness, this vibrant "web," itself a fundamental implication and property of a holographic matter-field.

The brain, as a reconstructive wave, is specifying past portions of the change of this field at a particular scale of time – a buzzing fly, or a heron-like fly, or spinning cylinder (once a rotating cube). The brain is establishing a certain *ratio* relative to the micro-events of the matter-field – in the fly's case, the micro-events making up the body of the fly, his wing beats, his internal processes. If we were to conceive of our body and the fly side by side within the matter-field at the null scale of time, we see there is no spatial differentiation between our body and the fly. Both of these "objects" are simply phases in the global transformation of this field. But allow the brain to gradually apply an increasing time-scale in its specification: the outlines of the fly begin to emerge, then the shimmering oscillations of his vibrant, organic crystalline structure, then slowly he begins to flap his wings, and then becomes the buzzing being of normal scale. The essential unity of the two – body and fly – within the matter-field is never broken. We arrive at Bergson's principle: subject is differentiating from object, not in terms of space, but of time.

But there is a simultaneous aspect in this specification. The body/brain, as a reconstructive wave, is simultaneously specific to, or specifying, from a given spatial perspective, a time-scaled form of the elementary web of awareness defined across the matter-field. Rotating cubes, stretching surfaces with gradients for rocks and grasses, little buzzing beings – are simply time-scaled forms of this awareness. So now we ask: Who is it that sees?

There is no one seeing. There is no homunculus viewing the image of the external world. The reconstructive wave, supported by body and brain, is simply specifying (specific to) a time-scaled portion of this vast web of awareness, this web of mind. This is why Douglas E. Harding found he "had no head," and that rather, he *was* the external world – the blue skies, the snow-peaks of the mountains, the misty blue-valleys of the Himalayas. These mountains, these valleys, these skies are Mind – Mind at a specific scale of time.

We begin life with the perception that we are the external world. This, I noted, was the observation of Jean Piaget, the great theorist of the child's cognitive development and one time student of Bergson (*The Construction of Reality in the Child*). The great constant in our experience is the body, and the body's spatial perspective. The body is the *invariant* in the matrix of experience. Our Identity gradually comes to settle upon it. We become an "object" among other "objects." We begin to need a koan.

Chapter II: End Notes and References

1. Gibson, J. J. (1975). Events are perceived but time is not. In J. T. Fraser & N. Laurence (Eds.), *The study of time II*. New York: Springer-Verlag, p. 299.
2. Robbins, S.E. (2004). On time, memory and dynamic form. *Consciousness and Cognition,* 13, 762-788.
3. Hummel, J.E. & Biederman, I. (1992). Dynamic binding in a neural network for shape recognition. *Psychological Reviews, 12*, 487-519.
4. Weiss, Y., Simoncelli, E., & Adelson, E. (2002). Motion illusions as optimal percepts. *Nature Neuroscience, 5*, 598-604.
5. Bergson, H. (1896/1912). *Matter and Memo*ry, pp. 6-7.
6. Bergson, H. (1896/1912). *Matter and Memory*, p. 28.
7. Bergson, H. (1896/1912). *Matter and Memory,* pp. 31-32.
8. Some examples:
   Viviani, P., and Stucchi, N. (1992). Biological movements look uniform: Evidence of motor-perceptual interactions. *Journal of Experimental Psychology: Human Perception and Performance*, *18*, 603-623.
   Viviani, P., and Mounoud, P. (1990). Perceptuo-motor compatibility in pursuit tracking of two-dimensional movements. *Journal of Motor Behavior*, *22,* 407-443.
   Churchland, P. S., Ramachandran, V. S., & Sejnowski, T. J. (1994). A critique of pure vision. In C. Koch and J. Davis (Eds.), *Large-scale Neuronal Theories of the Brain.* Cambridge: MIT Press.
9. Turvey, M. (1977). Preliminaries to a theory of action with references to vision. In R.E. Shaw & J. Bransford (Eds.), *Perceiving, Acting and Knowing*, New Jersey: Erlbaum.
10. Robbins, S. E. (2001). Bergson's virtual action. In A. Riegler, M. Peschl, K. Edlinger, & G. Fleck (Eds.), *Virtual Reality: Philosophical Issues, Cognitive Foundations, Technological Implications.* Frankfurt: Peter Lang Verlag.
11. Hameroff, S. and Penrose, R. (1996). Conscious events as orchestrated space-time selections. *Journal of Consciousness Studies, 3*, 36-53.
    Here "qualia" are virtually little atoms, residing in the microtubules of the neurons, just waiting to be configured into forms/experiences of the external world.
12. Bergson, H. (1896). *Matter and Memory,* p. 276.
13. Bergson, H. (1896). *Matter and Memory*, p. 255.
14. Bergson, H. (1896). *Matter and Memory*, p. 254.
15. Bergson, H. (1896). *Matter and Memory*, p. 263.
16. Bergson, H. (1896). *Matter and Memory,* p. 266.
17. Bergson, H. (1889). *Time and Free Will: An Essay on the Immediate Data of Consciousness.* London: George Allen and Unwin Ltd, p. 111.
18. Nottale, L. (1996). Scale relativity and fractal space-time: applications to quantum physics, cosmology and chaotic systems. *Chaos, Solitons and Fractals*, *7*, 877-938.
19. Lynds, P. (2003). Time and classical and quantum mechanics: Indeterminacy versus discontinuity. *Foundations of Physics Letters*, *16*, 343-355.
    A testimony to the sad, in my opinion, inexcusable burial of Bergson here: Not one reviewer of Lynds, to include J.J.C. Smart, had the thought to suggest that Lynds at least note his relation to Bergson's concepts.

20. Hardcastle, V. G. (1995). *Locating consciousness*. Philadelphia: John Benjamins, p. 1.
21. Robbins, S.E. (2010). *The Case for Qualia*: A Review. *Journal of Mind and Behavior* 31: 141-156.
22. Nida-Rűmelim, M. (2008), *Phenomenal Character and the Transparency of Experience,* in Edmond Wright (Ed.) , *The Case for Qualia.* Cambridge, Massachusetts: MIT Press.
23. Ramachandran, V. S., & Blakeslee, S. (1998). *Phantoms in the brain*. New York: Morrow.
24. Smythies, J. (2002). Comment on Crook's "Intertheoretic Identification and Mind-Brain Reductionism." *Journal of Mind and Behavior*, 23, 245-248.
25. Yarrow, K., Haggard, P., Heal,, R., Brown, P., & Rothwell, J. C. (2001). Illusory perceptions of space and time preserve cross-saccadic perceptual continuity. *Nature*, 414, 302-304.
26. O'Regan, J. Kevin (1992). Solving the real mysteries of perception: The world as an outside memory. *Canadian Journal of Psychology, 46*(3), 461-488.
27. Robbins, S. E. (2004). Virtual action: O'Regan and Noë meet Bergson. *Behavioral and Brain Sciences, 27*, 907-908.
28. O'Regan, J. K. & Noë, A. (2001). A sensori-motor account of vision and visual consciousness. *Behavioral and Brain Sciences*, 24(5).
29. Kovacs, I., Papthomas, R. V., Yang, M. & Feher, A. (1996). When the brain changes its mind; Interocular grouping during retinal rivalry. *Proceedings of the National Academy of Sciences, USA*, 94, 508-511.
30. Smythies, 2002, p.248.
31. Rock, I. (1984). *Perception.* New York: Scientific American Books Inc.
32. Robbins, S. E. (2007). Time, form and the limits of qualia. *Journal of Mind and Behavior*, 28, 19-43.
33. Bergson, H. (1896). *Matter and Memory*, p. 268.
34. Zeki, S. (1993). *A Vision of the Brain.* Oxford: Blackwell.
35. Pirsig, Robert. *Zen and the Art of Motorcycle Maintenance*, p. 213.
36. Bergson, H. (1889). *Time and Free Will*, pp. 100-101.

CHAPTER III

Retrieving Experience from the Holographic Field

...suppose that my speech had been lasting for years, since the first awakening of my consciousness, that it had been carried on in one single sentence, and my consciousness... able to employ itself entirely in embracing the total meaning of the sentence: then I should no more seek the explanation of the integral preservation of this entire past than I seek the explanation of the preservation of the three first syllables of "conversation" when I pronounce the last syllable. Well I believe that our whole psychical existence is something just like this single sentence, continued since the awakening of consciousness, interspersed with commas, but never broken by full stops.

- Bergson, *Mind Energy*

We cannot see how memory could settle within matter; but we do clearly understand how – according to the profound saying of a contemporary philosopher – materiality begets oblivion.

- Bergson, *Matter and Memory*

The Brain as Suitcase

The brain is considered, universally, to be the great storehouse of our past. It is our suitcase for carting around the memories of our experience. This is a scientific dogma. But in truth it is only an hypothesis. It has never been proven. The fact that a konk on the head can cause the loss of memories does not allow us to infer, categorically, that the memories, sitting in some storage area in the brain, have been destroyed. It does not allow us to rule out the possibility that the "konk" has simply damaged the mechanisms and dynamics necessary for accessing and retrieving experience.

Cognitive science, neuroscience, the psychology of memory, and last but not least, robotics – none have a clue how experience is actually stored in the brain. The prevailing trend has at least partial origins in the "constructivist" tradition of Ulrich Neisser (circa 1967) where memories are apparently reconstructed from vaguely specified "pieces" or elements. This is reinforced by an "abstractionist" tendency. The abstractionist sees only certain abstracted aspects of the events of experience being stored. The rotating cube would have only certain snapshots of the rotation stored. For the memory theorist, Lawrence Barsalou, the dynamic transformation of "biting" as in biting on a carrot, would be represented by three schematic states – "a mouth closed next to the object, followed by a mouth open, and then the mouth around the object" (Figure 3.1).[1]

Figure 3.1. "Biting" as a schematic abstraction.

There is no principled theory as to how or why the brain makes such a selection. *Which* snapshots of the rotating cube are stored? (We have seen that in actuality there can be no such snapshots as far as the brain is concerned.) *Which* snapshots of the biting face? In fact, there would be a time-extended flow field defining this biting transformation just as difficult to store as the flowing sides of the rotating cube. Yet, say theorists Mayes and Roberts, "Only a tiny fraction of experienced episodes are put into long term storage, and, even with those that are, only a small proportion of the experienced episode is later retrievable."[2] It is from these abstracted elements, the principles of selection of which no one can specify, that the brain is to reconstruct specific experiences.

This ubiquitous view is intended to alleviate the burden on the brain of storing the vast volume of our experience, as well as eliminate the burden of how the immensely rich *images* of these experiences could be stored. One only needs to get up and walk outside for five minutes, observing the vastness of visual detail presented in experience to feel sympathetic for this position (if you are a storage person). However there are only vague notions on how the brain makes this selection of elements or events, or for any particular event, which elements these would be, or how dynamic and multi-modal events are then reconstructed from fragmented pieces or static "features" of an event – as for example the ephemeral edges and vertices of the rotating cube which we saw exist only as invariants over a flow – are stored in different (unspecified) spots, frozen, as snapshots, in the brain. All theory proceeds as if it is sufficient to theorize on how the hippocampus (and/or other structures) might store some static event, e.g., a snapshot or series of snapshots of a person with a spoon stirring our cup of coffee.

Let us remember, in the previous chapter we saw that there is no theory, in fact, not even a teensy thought of a theory, for what I termed the "primary" memory that supports the time-extended perception of the rotating cube or the buzzing fly, let alone, as Barsalou's weak attempt shows, of storing these dynamic, time-extended events for future retrieval. Let us remember too, the question of whether the brain stores experience was Bergson's key question in determining whether the brain generates consciousness. If we hold that neural storage is sufficient to account for the retention of experience, it must be understood that this answer absolutely constrains all theories of the origin and nature of consciousness. Images, dreams, even perceptual images and perceptual experience, to include the time-extended perception of the rotating cube, *must* somehow be generated from stored elements within (or modifications of) the neural substrate. But we have seen that the metaphysic underlying this position, with its abstract homogeneous continuum and its snapshot theory of motion, is utterly, logically precluded from explaining the origin of any of these. The fact is, I repeat again, current theory has not one clue how experience is stored in the brain. Of course, it has not one clue what experience, that is to say, perception, is. Without a theory of perception, there can be no theory of the storage of experience. This is why there is no theory.

Nevertheless, despite this extreme theoretical weakness, the notion that the brain, and therefore a robot, can store experience, is a lynchpin, both in theories of consciousness, and in the robotic, machine view of mind with its view that robots can be conscious, with the philosophic mist that this generates. If the brain cannot and does not store experience, then neither will a robot, a computer or a neural network.

The Brain as Not a Suitcase

The dynamically transforming visual image of the external world, we have seen, *is not within the brain*. It is right where is says it is, in the external field. Therefore it cannot be stored within the brain. Experience, in general, cannot be stored there. The material brain is the three-dimensional cross-section of a four-dimensional being, flowing in time. Bergson visualized the brain as a sort of "valve," holding back the accumulated waters of experience. When the systems in the brain for preparing action momentarily configure in a certain manner, the brain/valve allows past experiences into consciousness that fit this action configuration. For example, driving down a certain section of curving road entails a certain form of the body's action and by this we may recall a similar driving experience. The loss of memories on the other hand – amnesias, aphasias – does not mean that the experiences are lost or their files destroyed in their cabinets in the brain. Rather, Bergson argued, only the ability of the brain to assume certain action configurations has been damaged.

In today's terms, we would say that in the recall of experience, the brain is yet acting as a modulated reconstructive wave passing through the four-dimensional holographic field. Just as we modulated the reconstructive wave to different frequencies to reconstruct the successive wave fronts of the pyramid/ball, cup, toy truck and candle, so it is the modulated wave patterns that the brain assumes which determine the *experiences* retrieved. Damage to the brain in various areas affects its ability to assume the complex wave-modulation patterns that might be required to retrieve certain experiences. The experiences now appear to be "lost." The brain does not have to be our "suitcase" for experience. What is required is that it serves to assume the appropriate modulated reconstructive wave patterns within the holographic field.

We should not take this to mean that no form or type of memory whatsoever is stored in the brain. To use a Bergson-like example, we may have many practice sessions at the piano, say, learning Chopin's C# minor waltz. The end result of the many practice sessions is a nice motor program – with all the modifications of neural connections this might involve – which we can unroll effortlessly as the waltz. This is clearly a form of memory "stored in the brain." In the terms of current psychology, it might be called a "procedural" memory – a memory for a sequence of actions. There are other neural storage effects. The layout of the piano keyboard, with its five black and seven white alternating keys, is experienced daily over and over. The particular black-white pattern is always the same; it is an *invariant* over these experiences. The spatial neural firing patterns in the visual cortex which respond to the layout of the keyboard must eventually be registered at some level of the neural structure or hierarchy as an invariant pattern. Jeff Hawkins (*On Intelligence*) proposes a neural network model for the formation of these all-important neural "invariants." Were a different keyboard with a different white-black alternation pattern encountered one day, the difference would be detected instantly. But simultaneously every experience, every practice session or episode is retained in the four-dimensional extension of being – the day when the sun was shining brightly and the room was glowing, the dreary day when it rained, the day we had a headache. This is "episodic" memory in current terms. In principle, given the precise modulation pattern, each event is retrievable. In practice this is obviously difficult.

The relation between the form of memory stored in the brain, and our experience, which is not so stored, is a complex one. Not all the aspects of it I have examined in other theoretical endeavors will be discussed here. But let us explore a little more concerning the fundamental retrieval mechanism of experience.

Retrieving Events – Redintegration

The fundamental operation of the retrieval of experience is "redintegration." This curious term, which I happen to like, can trace its genes back to the very origins of psychology. Thus it was Christian Wolff, a contemporary and disciple of the great philosopher Leibniz, and a mathematics professor (it took one to coin this term), who first introduced the "law of redintegration" in his *Psychologia Empirica* of 1732. In effect, Wolff's law stated that "when a present perception forms a part of a past perception, the whole past perception tends to reinstate itself."

Examples of this phenomenon abound in everyday experience. Thus the sound of thunder may serve to redintegrate a childhood memory of the day one's house was struck by lightning. Perhaps, for example, we are walking down a road in the summertime and suddenly notice a slight rustling or motion in the grass along the embankment. Immediately, an experience returns in which a snake was encountered in a similar situation. Klein (*A History of Scientific Psychology*) notes that these remembered experiences are "structured or organized events or clusters of patterned, integrated impressions," and that Wolff had in effect noted that subsequent to the establishment of such patterns, the pattern might be recalled by reinstatement of a part of the original pattern.

One would think that the science of describing these "patterns" precisely, such that we can predict when retrievals of experience will occur, would be seen as crucial. Such a science was in fact provided by Gibson in his invariants, invariance laws and gradients. It is a gift that the psychology of memory and cognitive science have for the most part ignored.

The "law" for describing when a present experience will retrieve or redintegrate a past experience is fairly simple: *When a present event has a set of invariance laws defining it, it will retrieve a past event with the same invariance law description.* In essence, when the same dynamic pattern, supporting the same invariance laws or structure, is evoked over the global state of the brain, the correspondent past *experience* is reconstructed.

Let us put this into our event context. Imagine a drive up a mountain road. The road curves back and forth, in sinusoidal fashion, rising at a particular grade. We have then a certain gradient of velocity vectors lawfully transforming as a function of the radius of the curves and the velocity of the vehicle. There are other components such as the contour and texture density gradients, for example the distribution of rocks peculiar to a mountain terrain. Over the flow field there is a ratio called tau (τ), already noted in the previous chapter (Figure 2.8). The essential point is that the tau ratio specifies time to impending contact and severity of impact. It has a critical value in controlling action. A bird or a pilot, I noted, coming in for a landing on the flowing surface below, uses tau to land gently. Our driver can rely directly on the tau value to modulate his velocity to avoid possible impacts with structures along the road. Thus the transformation specifying the flow field also contains information for "tuning" or guiding the action systems. The velocity of field expansion/directional change is specific to the velocity of the car and to the muscular adjustments necessary to hold it on the road. Therefore the state of the body/brain with respect to future possible (virtual) action as well as that actually being carried out constitutes an integral component of the event pattern.

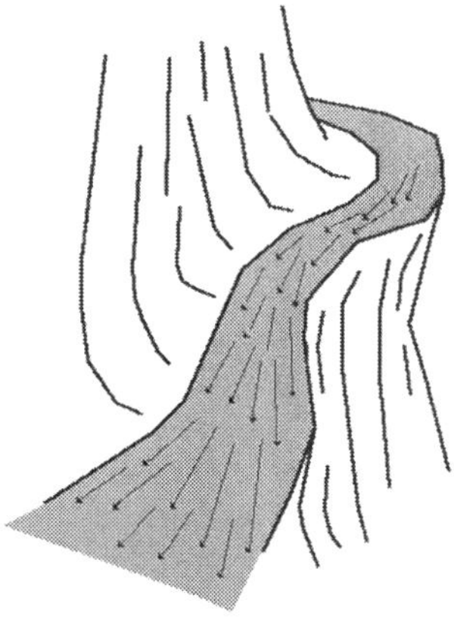

Figure 3.2. The mountainside drive

I believe it is quite common for people to have past experiences redintegrated by this form of flowing, road-driving, invariance law structure. My wife tells me that every time we drive along a certain curving section of the freeway near Milwaukee the memory pops back of driving on a segment of a freeway in

California where she once lived. Due to the indivisible time-evolution of the matter-field, neither the original transformation of the field nor its subsequent specification by the brain as an image (of its then-past motion) have moved into non-existence. The reason that the experience is reconstructed is that the brain is thrown by the invariance structure of the present event into the same reconstructive wave pattern as that which defined the original event.

This is the essence of the principle of redintegration.[3]

Nowhere in the description of this process, one will note, is it necessary to assume that the experience being reconstructed is stored in the brain. In truth, I noted earlier, cognitive psychology has no idea how experience is actually stored in the brain, nor does computer science, nor does robotics.

Let us consider one other event: let's make it "stirring coffee in a cup, using a spoon." This event has a time-extended invariance structure. Thus, the swirling coffee surface is a radial flow field. The constant size of the cup, should it move forward or backward, is specified, over time, by a constant ratio of height to the occluded texture units of the table surface gradient. There is again a tau ratio defined over this flow field; it supports modulating the hand for grasping the cup. Were the cup cubical, its edges and vertices are sharp discontinuities in the velocity flows of its sides as the eyes saccade, where these flows specify the form of the cup. The periodic motion of the spoon is a "haptic" (meaning the information in touch) flow field. This haptic field carries what physics terms an "adiabatic" invariance – a constant ratio of the energy of oscillation to the frequency of oscillation. The action of *wielding* the spoon is defined by an "inertial tensor." This is a matrix whose values describe the spoon's resistance to angular acceleration. This is but a partial set of the invariance laws defining this little event.

Before we go on, let me define an invariance structure: An invariance structure is *the set of transformations and invariants specifying an event and rendering it a virtual action*.

Retrieving Events 2

We have seen (Chapter 1, Figure 1.4) that when we pass a reconstructive wave through a hologram, we can modulate the wave to different frequencies, for example, to f_1 to specify the wave front of a pyramid/ball, to f_2 to specify the cup, to f_3 for the toy truck, etc. The more spread apart (or unique) the frequencies are, the easier it is to modulate the wave to each precise frequency, and the easier it is to reconstruct each separate wave front. If f_1 and f_2 are very close, it may be very possible that a reconstructive wave has both frequency components, f_1 and f_2, in which case we get a confused, composite image of both the pyramid/ball and the cup. In this case it turns out to be difficult to reconstruct the individual "events" (wave fronts). The easier it is to create a unique reconstructive wave, the easier it is for precise, individual recall.

The brain is dealing with "events" (or experiences). It is the structure of events, in terms of the invariance laws defining an event, that drives the modulation of the brain's reconstructive wave. If we need to remember a *set* of events, then the more unique the structure of each event, the easier it is for the brain to achieve the unique modulated reconstructive wave required to reconstruct each event.

This is a simple principle, but memory research has mostly danced around it. This is largely due to, a) ignoring Gibson, and, b) focusing on *verbal* materials so heavily in the experiments that the ecological nature of events is obscured. I am going to spend just a moment relating our brain-as-reconstructive-wave model to the standard literature of memory research.

The Verbal Research

In the 1970's, memory research was dominated by experiments using some form of verbal material. The initial paradigm was introduced by Herman Ebbinghaus in the 1890's. Ebbinghaus introduced the use of "nonsense syllables." He experimented with learning and remembering lists of syllables like QEZ, PUJ, KAX, etc. Another form involved lists of pairs like QEZ-PUJ or ZAK-WUX. When presented QEZ again, you had to remember that PUJ was its pair. It was therefore Ebbinghaus's attempt to remove all *semantics* from the memory task. He had found it too easy to remember pairs like GRASS-SNAKE or SPOON-COFFEE due to the "associations" already existing in his past experience. This was the drive to get at the formation of the elementary "item-bond," e.g., between QEZ and PUJ, in its formation. In essence, the attempt was being made to drive the human "device" as a far lower version of itself, what in computer terms would be called a "syntax-directed" processor. The lists of QEZ-PUJ-like pairs were sets of arbitrary syntax rules, to be laboriously associated. This was to be the primitive process of association on pure display.

My experience, in studying the vast literature on all these experiments, was one of walking deeper and deeper into a vast realm of trees grown and even sprouting everywhere. There seemed no view possible of the forest.

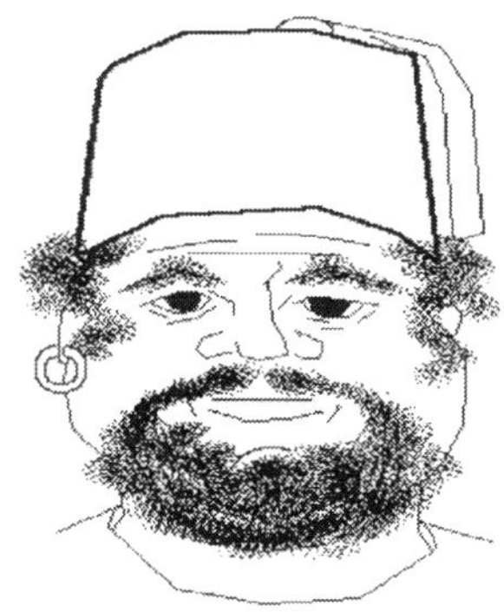

Figure 3.3. QEZ-PUJ – the nonsense syllables transformed to a pudgy Turkish person with fez.

Despite the apparent purity of the experimental verbal materials, there really was no experimental control at all over what the subject did in his head. He was free to use imagery such as the Turkish person, or his experience of snakes, and so on. Soon the research field found itself having to deal with troublesome "variables" such as "meaningfulness," and "context" in the theoretical models of the process. The mini-models of the "process" and laws of memory began to get more and more complicated and difficult to grasp. There was no sense at all that the ecological case, using real, concrete events for remembering, might be primary, or that in fact, it is in the ecological case where the greatest experimental control resides.

But as history would show, the "device" was far more powerful, and its true (holographic) properties kept intruding into these pure associationist experiments in a pesky, unwanted way. A subject might realize, awakening from his rote efforts, that QEZ could be seen as "FEZ," and PUJ stand for "pudgy," and thus become a pudgy Turkish person wearing a fez (Figure 3.3). Now he had a nice method for redintegration using the powers of his "device" the next time QEZ (now "FEZ") came along. In my graduate student days, in the 1970's, there were still tons of nonsense syllable experiments being performed (and they continue a bit today). A standard experiment in the verbal learning tradition of memory research has the experimental subjects learn a "paired-associate" list. Below is a very short list (they usually run about 10-20 pairs). The goal is to learn that when one sees SPOON, that the response or answer is COFFEE, or when one sees KNIFE, the response or answer is SOAP. All the pairs (first "spoon," then "coffee") are presented one by one. It may take several complete presentations of the list to learn all the pairs.

List 1

SPOON-COFFEE
KNIFE-SOAP
BOTTLE-THIMBLE
And so on....

In the course of years of experiments, the daring discovery was made that if the subjects formed *images* of events for the pairs, learning was greatly sped up.[4] For example, we imagine the SPOON stirring the COFFEE, or the

KNIFE cutting flakes from the SOAP. In other words we get a little more ecological, for now we are actually remembering *events*, not just arbitrary word pairs. From the point of view of experimental control, this is not all that satisfactory as far as discovering the laws of memory – we have no true clue what events the subjects are imagining.

A more complex problem for learning word-pairs involves *two* lists.

List 1 (A-B)	List 2 (A-C)
SPOON-COFFEE	SPOON-BATTER
KNIFE-SOAP	KNIFE-DOUGH
BOTTLE-THIMBLE	BOTTLE-PAN

This is the "A-B, A-C" paradigm. Here the task is to learn the pairs of List 1, then turn right around and learn the pairs of List 2. But this is no easy task, since, if you notice, the "A" portion of the pairs is always the same, e.g., **SPOON**-COFFEE (list 1) and **SPOON**-BATTER (list 2). This leads to "interference," for when we see SPOON, what word will it redintegrate – COFFEE or BATTER? The response word from List 1 interferes with learning the word in List 2, and vice versa. This interference has led to countless theories as to what it is, how the word-pairs are stored in the brain, the nature of the memory "trace" in the brain (no one knows what a "trace" actually is), and so on.

I am going to ignore all the theoretical entanglements and go right to the ecological case. Let us assume these are *concrete events*, enacted or perceived. The subject stirs the coffee with the spoon, or stirs the batter with the spoon. He cuts the soap with the knife, or cuts the dough. He pours water from the bottle into the pan, or into the thimble. Now in the verbal case, after learning the lists, we would simply present the word "SPOON" as a cue. But this is a very vague event; it has no *specificity*. Which object, or which event will it redintegrate? Even a concrete but static spoon, placed on the table before us, would have a questionable cueing power. We have at best the classic "response competition" model which was introduced by McGeoch, a theorist in 1942 (Figure 3.4). In the behaviorist terminology of the day, the "stimulus,"

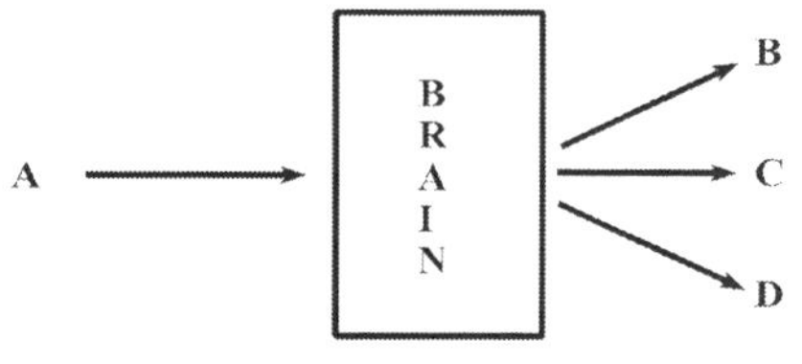

Figure 3.4. McGeoch's Interference Model (1942). Event A activates or reconstructs events B, C and D, or "responses" B, C and D. These events, B, C, D interfere with each other and make retrieving the desired memory difficult.

SPOON, activates both BATTER and COFFEE as possible "responses." In modern terminology, we would say both words are "primed." In essence, we could say, we have sent a highly *unconstrained* wave though our holographic memory, as though we had a non-coherent wave containing frequencies f_1 and f_2, the frequencies of the original reference waves for recording two different wave fronts (or objects).

We can improve this by modulating the brain's reconstructive wave to a precise or coherent frequency, correspondent with the original reference wave, e.g., f_1, of the desired stored wave front. And concretely, in terms of events, we must put greater *constraints* on the cue event. For starters, a dynamism must be placed on the static spoon – the circular motion of stirring. But yet more precise constraint is needed to cue the proper event and object – batter or coffee. We can choose to create the greater resistance provided by the batter medium, or the larger amplitude of circular motion in the batter case assuming it was in a bowl, or both. The cue-event must force this precise dynamics and therefore modulatory pattern in the brain. Had one of the pairs been SPOON-OATMEAL, where the spoon was being used to shovel-in the oatmeal, a quite different transformation, involving scooping and lifting, and with the weight, mass and consistency constraints implied by oatmeal, would be placed on the cue-event with its spoon.

With this in mind, one can imagine an impossibly difficult paired-associate paradigm as far as verbal learning experiments are concerned. We'll call it the "A-B_i" paradigm. A list would look as follows:

A B_i
SPOON-COFFEE
SPOON-BATTER
SPOON-OATMEAL
SPOON-BUTTER
SPOON-CORNFLAKES
SPOON-PEASOUP
SPOON-CATAPULT
SPOON-CHEESE
SPOON-TEETER TOTTER
And so on…

It is "impossible" since, in the purely verbal case, the stimulus words are exactly the same and the subject could have no clue what the appropriate response word is. But assume that the subject concretely acts out, while blindfolded, each event of the "absurd" list – stirring the coffee, stirring the batter, scooping/lifting the oatmeal, the cornflakes, cutting the cheese. To effectively cue the remembering of what paired object was involved, as I have said, the dynamics of each cue-event must be unique, and the structure of invariance laws of an event in effect implies a structure of constraints.

These constraints may be "parametrically" varied, where increasing fidelity to the original structure of constraints of a given event corresponds to a finer tuning of the reconstructive wave. The (for example, blindfolded) subject may wield a stirrer in a circular motion within a liquid. The resistance of the liquid (a parameter value) may be appropriate to a thin liquid such as coffee or to a thicker medium such as the batter. The circular motion (a parameter value) may be appropriate to the spatial constraint defined by a cup or to the larger amplitude allowed by a bowl. The periodic (back and forth) motion of the hand may conform to the original "adiabatic" invariance (the ratio of frequency/energy) within the event, or may diverge. We can predict that with sufficiently precise transformations and constraints on the motion of the spoon (either visual, or auditory or kinesthetic or combined), the entire list can be reconstructed, i.e., each event and associated response word. Each appropriately constrained cue-event corresponds to a precise modulation (or constraint) of the reconstructive wave defined over the brain.

The obvious inverse is that, as the parameter values diverge from the original event, cueing/recall performance and/or recognition performance will increasingly degrade. Recognition tests are one method of testing these manipulations, employing familiarity ratings. In this case, we would present, as recognition cues, the original events transformed on various dimensions. Experimental subjects rate the re-presented event with a value reflecting how sure they are that they have previously seen it (how familiar). The familiarity value should steadily decrease as the parameters are varied away from their original values.

Parametric Variation of Memory Cues in Concrete Events

These kinds of experiments are implicit in the literature. They are just not realized as "fits" in the theoretical structure I have just described. An example is found in a demonstration by Jenkins, Wald and Pittenger.[5] Capitalizing on the notion of the optical flow field, they showed subjects a series of slides that had been taken at fixed intervals as a cameraman walked across the university campus mall (Figure 3.5). Some slides, however, were purposely left out. Later, when subjects were shown various slides again and asked if they had seen the slide shown, they rejected easily any slide taken from a different perspective and which therefore did not share the same flow field invariant defined across the series. Slides not originally seen, but which fit the series with its flow field were accepted as "having been seen" with high probability. But Jenkins et al. had created a "gap" in the original set shown to the subjects by leaving out a series of six continuous slides. Thus a portion of the transformation of the flow field was not specified. Subjects were quite easily able to identify these slides as "not seen." In this case, we are in effect varying parametric values defining a flow field. We are seeing that redintegration, and therefore *the brain's retrieval dynamics*, is indeed sensitive to this form of parameter manipulation. Other manipulations are possible, for example the slant of the gradient, the smoothness of the flow, the velocity of the flow, etc.

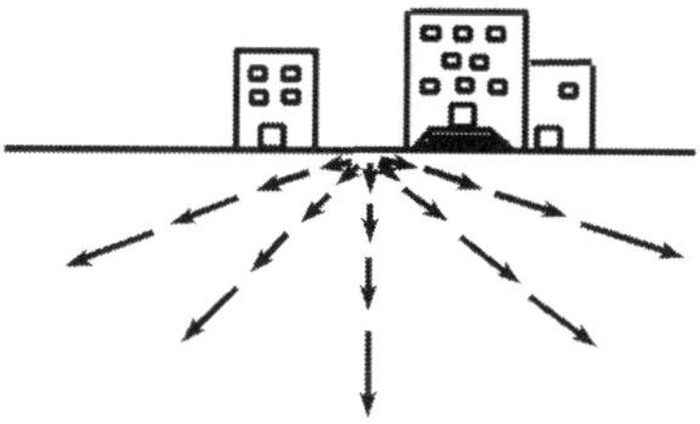

Figure 3.5. A flow field is created while moving along the campus mall

We can consider, for example, presenting a simple static event of a field with a schematic tree (Figure 3.6). The trees in the figure have been grown with the precise mathematics defining real tree growth.[6] The number of terminal branches (N) is a set function of height (H), $N=k(H)^2$, while the diameter (D) of trunk or branches at any point is a function of remaining length to the tip, $D=k(H)$. Suppose that one of the recognition items in our experiment is the leftmost or youngish tree in Figure 3.6. In the recognition test phase, the parametric values defining the structure of the trees can be varied increasingly from the original value. For example, we could re-represent the second tree, or third, or fourth tree and ask if this is recognized as part of the original set of items. Familiarity ratings should drop the further we move along the age dimension. More dynamically, a time-accelerated view of the tree's growth under certain parametric values can be used as the stimulus. For a dynamic event such as an approaching rugby ball, another value defined over the velocity flow of the approaching ball known as "time to contact" (which specifies the immanence of the object hitting us) can be varied increasingly from the original value.[7]

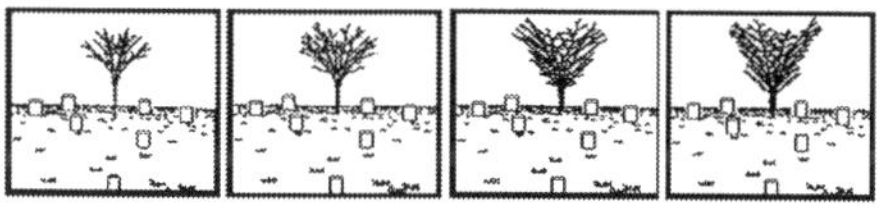

Figure 3.6. A generated, aging tree. (Adopted from Bingham, 1993)

Even what for us is a very slow event, namely, the aging of the facial profile, has its invariance structure. Pittenger and Shaw, we saw, showed that this event can described and generated via a strain transformation on a cardioid (Figure 2.10). Their original experiments can be re-cast in this redintegration test framework.[8] Originally the subjects looked at many pairs of faces, each face being generated by this law with varying values of the strain transformation, and judged each time which of the pair was the older. Changing this to a memory task, a face of a certain age can be included in a set of various items successively presented to a subject. On the recognition task, a face transformed by a certain parametric aging value is now presented. Familiarity values will be a function of the closeness to the transformation or strain value. The aging transformation works for animal faces too, even for Volkswagens – it can generate increasingly aging "Beetles!" So we can have many different kinds of items in this test that eventually get aged (or un-aged) in a recognition phase.

All these are memory experiments that are either waiting to be done or implicitly have been done.[9] They demonstrate that memory and its supporting brain dynamics are extremely sensitive to the invariance structure of events and the actual "parameter" values of the transformations involved. The importance of this is that any theory that claims to be a model of memory must be able to support this dynamic structure in events. But as we shall soon see, the major claimant to the theory of memory, namely neural networks or "connectionist" networks, is utterly incapable of this.

The concrete ecological event, which we have focused on here, is primary in the theory of memory. This is the real world in which the brain and body were designed to concretely operate – to avoid an unfriendly tyrannosaurus (they were probably all unfriendly), to spot a saber tooth tiger, to survive. Curiously, after perhaps forty years of memory research in the abstract verbal world, a small core of researchers based primarily in Sweden and Norway began, in the 1980's, to study memory in the ecological context.[10] They now study what is termed "Subject Performed Tasks" (SPTs), where subjects concretely act out lists of commands such as "bend the wire," or "break the match." The subjects are then asked to recall the series of actions. Not surprisingly, the results of the experiments call into question many sacred cows of the verbal learning research. Memory performance for example is vastly superior to pure verbal list learning, being often near perfect for the acted-out lists of 20-30 events. In other words, there is little "interference" among these events.

But the SPT group, along with the rest of cognitive psychology, has little in the way of a model as to how these experiences are "stored" or how or why actions help retrieve the experience. Speculating on the mechanism behind the free recall of such acted events in SPTs, Zimmer et al. use the notion of "popping into mind," as in cued recall, where "sets of features," "bound together by actions," can pop out from the noise if their "conjunctions" are sufficiently unique (p. 669).[11] These "spontaneously reconstitute" the former episode. This is little improvement, however, albeit with a greater appreciation of action, beyond Klein's reformulation of Wolff's statement of 1732, and it has no theory as to how these static features are assembled to become an event or event-image. Seizing only on the "acting out" of the event as the critical difference between SPTs and the standard verbal materials, the SPT group is a bit flummoxed by the near equal memory performance for EPTs, or Experimenter Performed Events, where it is the experimenter, not the subject, that "breaks the match" or "bends the pencil" and the subjects simply observe. They fail to see that the EPT is providing as near a full specification of the invariance structure of each event as is the SPT, with nothing left to the subject's imagination. This blindness to the invariance structure of events surfaces in the attempts of the SPT researchers to have subjects associate verbs to verbs (actions to arbitrary actions) in SPTs, e.g., "lifting the book" as a cue for "tapping the wall." The subjects were not, of course, successful.[12] There is nothing in the invariance structure of the first capable of redintegrating the second. This was in effect a nonsense syllable task imported into the context of SPTs.

Standard memory theory likes to say that it is the "distinctiveness" of events that makes them easier to remember. The distinctiveness might be a function of "sets of features" of the event. The more "distinctive" events A, B and C are, the easier it is to "encode," even "specifically encode" these events in "storage," and the easier it is to cue each event for later recall. Though there is no clue what this encoding could mean for events like stirring, theory has never gone beyond this vague equation or reference to event "features." Despite the fact that this distinctiveness/encoding formulation again represents not one step beyond Wolff's statement of 1732, memory theorists question the need for casting memory theory in terms of the invariance structure of events. They ignore that it is the precision of control of an event for modulating memory retrieval, in this case controlling a cue-event, that is the name of the game in science. And this control can only be through knowing the invariance laws or structure of an event.

To add to the reluctance, theorists note the difficulty that would exist in defining these invariance structures. Unfortunately, discovering invariance laws is also the name of the game in science. It also shows, to the chagrin of the memory theorists, that memory theory cannot be divorced from perception.

Some other already studied event contexts in the ecological literature to which this parameterization of dynamic structure is extendable: the severity of contact between two approaching objects,[13] volume related to perceived heaviness,[14] acoustic information for the motion of objects,[15] spoon handling in children,[16] perceived relative mass in physical collisions,[17] time-to-topple (say, of a bottle)[18] walk-on-able slopes,[19] sensitivity to the relation of period and length in pendulum motion,[20] acoustical frequency in vessel filling (where sound pitch increases as a vessel is filled with liquid).[21]

Remembering Baseball Games

We have begun with very concrete, ecological events. This is where things start. But two memory researchers, Vicente and Wang, saw that the same principles of memory hold for complex tasks in which levels of *expertise* can be developed, tasks such as chess, computer programming, industrial models of power plants, baseball, medical diagnosis, etc.[22] In this context, a wide range of studies has shown an *expertise effect* in memory. In chess, early work by Adrian de Groot asked four players of varying skill level to reconstruct meaningful board positions after only a few seconds of exposure. The Master and Grandmaster level of players reconstructed the boards with near-perfect accuracy, while the performance of the lesser players was less impressive. It would later be discovered that if the board positions consisted of randomly placed pieces, the performance of these experts dropped to a level insignificantly different from the rank novice.

What is the source of the expertise effect in memory? The answer is that there are expertise effects when there are constraints that experts can exploit to structure the events they are perceiving. The more constraint available, the greater the advantage. In the chess example, the constraints determining possible patterns of chess positions are set by the rules of the game. These had been removed when the positions were random. The structure had been destroyed. The degree of the effect is proportional to the number of constraints removed. When there are no goal relevant constraints present in the event to be remembered, the expert is reduced to the novice.

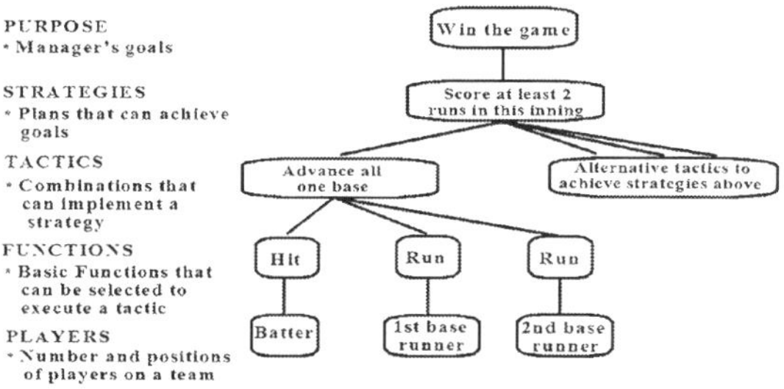

Figure 3.7. An abstraction hierarchy for baseball

To represent the constraints in these complex domains, Vicente and Wang would adopt a representation method termed an *abstraction hierarchy*. This is a stratified hierarchy where each level below the top level describes the means to achieve the ends at the higher level. At the top level is the purpose or goal. A very simplified example applying to baseball, specifically from the perspective of a decision making manager, is reproduced in Figure 3.7.

Imagine trying to remember this baseball vignette:

> ...The Senators are at bat with three outs and two pitchers, Smith and Jones, on first. Whitcomb is on deck. He knows that the next batter is their weakest hitter, so Whitcomb's goal is to throw a strike to load the bases. The shortstop looks towards the outfield and home for wide lead-offs by the fielders, then throws a pitch for a strike.

This is an example of an event description with a good number of violations of constraints, or equivalently, violations of normal baseball event invariants, and therefore presenting greater difficulty for memory for the baseball expert.

We are all experts on coffee stirring. With the exception of the newborn, very young, or visitors from Mars, these kinds of elemental, everyday events have a structure that is understood. It is an *experiential* expertise, not that of an engineer or physicist viewing the coffee cup. Describing these as invariance structures is the most appropriate approach. Baseball itself consists of numerous sub-events that are invariance structures – hitting the ball, catching the ball, throwing the ball, running, standing around, etc. These themselves become larger invariance structures – hitting the ball and running to first, catching the ball and throwing it to the infield, etc., and the abstraction hierarchy has attempted to capture the whole of this as a system of constraints relative to these events as a *game*.

The expert baseball fan sees the events playing out before him or her on the baseball field through this "lens" of invariance laws. It is a dynamic structure of perception gained only over concrete experience of playing and observing the game. The expert computer programmer likewise looks at system diagrams, program code structure, even core-memory dumps of problem programs through such a similar, dynamic "lens" of invariance. As we shall see, this increasingly complex perception is the growth of a *skill*. It is a skill based in something that the computer model of mind will never support – the invariance laws of concrete events.

Priming the Memory Pump

"Priming" is considered to have major implications for the structure of memory. We need to look at it a bit as it plays into the notion of implicit (unconscious) and explicit (conscious) memory which we will look at in the next chapter.

In one experimental paradigm a word is presented briefly, e.g., "spoon." This is followed quickly by another word or non-word, as an example in the word case, "coffee." The subject's task is usually a simple one like indicating whether "coffee" is a word or a non-word. In the SPOON-COFFEE pair, there would be an expected priming effect since SPOON is a close "associate" of COFFEE and has prepared the way in some sense for the response. In a pair like SPOON-BOOK we might expect a lesser priming effect as BOOK is not a close associate of SPOON. The effect is measured by response times. BOOK, because it is primed less well, would require a longer response time by the subject to indicate, via a button press, "word."

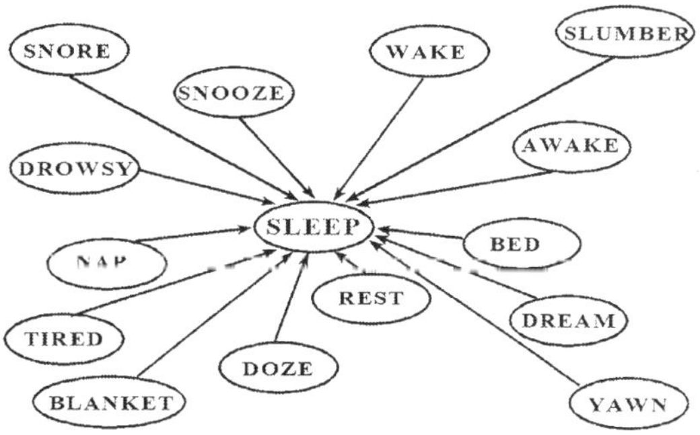

Figure 3.8.. A semantic activation network for the concept "sleep." The words shown are the fifteen highest "associates" to sleep.23

A major theoretical explanation for the priming effect has been the "spreading activation" model. Here memory is conceived as a network of "nodes" consisting of concepts related by semantic links or associative links. Consider the concept of SLEEP with its semantic network (Figure 3.8). In the classic, spreading activation-explained phenomenon, the subjects hear fifteen of the surrounding concepts/words (bed, rest, awake, tired, etc.). Though "sleep" is not presented, on a later recognition test, the subjects are extremely likely to "recognize" it (wrongly) as having been part of the original list they heard.[24] "Sleep" has been (unconsciously) "primed."

"Stirring" is no less a dynamic invariance structure than "sleep." On seeing SPOON, activation would be conceived to spread through the network of nodes related to SPOON, ultimately reaching COFFEE and facilitating response time to COFFEE. But what is stored at these nodes? If our node is STIR, is it the dynamic, time-extended, multi-modal invariance structure we will shortly see described for a coffee stirring event with its invariant visual radial flow field, its invariant forces for the "wielding" of the spoon in a stirring motion, the resistance of the liquid, the auditory "clinking" and more? But how is this invariance, which has no reality other than as an invariant defined across concrete *experience*, stored "at a node?"

Another competing model is the "compound cue" theory. Here the SPOON-COFFEE pair is viewed as forming a compound cue which is used to create a match to information in long-term memory. This joint-cue is a more powerful cue, providing the basis for a familiarity or parallel matching process to all items in memory. Facilitation of a response then is considered a function of "familiarity." A pair like MOTHER-CHILD is more familiar than MOTHER-HOSPITAL. The familiarity value is conceived to relate directly to the response time required to categorize CHILD or HOSPITAL (as a word or non-word).

The SPOON-COFFEE pair is really but a compacted version of an experimental paradigm where *sentences* are presented such as:

(1) The spoon stirred the [COFFEE].
or,
(2) The spoon stirred the [BOOK].

There is again a quicker response time in (word/non-word) categorizing COFFEE in (1) than BOOK in (2). Again, (1) is presumed to have greater "familiarity" than (2), enhancing the response process. However the usefulness of this type of experimental material has been rejected because researchers saw no way to get "familiarity" values from the sentences, whereas with single words they could use associative frequency norms, e.g., the frequency that MOTHER appears as an associate to CHILD as opposed to HOSPITAL, or SPOON with COFFEE as opposed to BOOK.[25]

Yet, with the sentence paradigm, we begin to approach the ecological case – real, concretely specified events. Again we have extremely little control over the events in the subject's mind when we present word pairs like SPOON-COFFEE or DEER-GRAIN. Suppose we had this set of sentences:

(3) He stirred the coffee with the [spoon].
(4) He stirred the coffee with the [knife].
(5) He stirred the coffee with the [orange peel].
(6) He stirred the coffee with the [truck].

Here we are clearly in the case of an event invariance structure. "Coffee stirring" specifies an equivalence class of objects that can participate in the event, and that can fill in the blank. We have sent, in effect, a reconstructive wave through memory defined by the constraints of the invariance structure. Sentence (3) is the level of invariance defining normal (global) context. It is most "familiar." Sentences (4), and (5) begin defining a dimension of possible substances and structures which can participate in the event given they support certain structural invariances. Sentence (6) sits way at the end of this dimension, if at all. Nevertheless, with proper (global) context, for example a pre-discussion of childhood play, I could likely bring up the response time on (6). "Familiarity" or networks of "associates" are only a poor approximation to describing the effect of the dynamic patterns of activity and invariance structures that actually exist in the four-dimensional memory of being.

Priming is another case of redintegration with its inherent reliance on event invariance structures. Once thought immune, it is now known to be fully subject to the same "interference" effects as in the old McGeoch model – as we would expect with redintegration.[26] This also means that priming is also subject to the parametric variation of these structures. This can be kept then in the very ecological dimension where it must be initially understood. Therefore we might have the equivalent of a "priming sentence" such as:

(7) The spoon stirred the ________ .

But let the event be concretely acted out, e.g., the (blindfolded) subject actually stirs with the spoon within standard spatial constraints, where the substance stirred has liquid properties, thus resistance, similar to coffee. Now we present words/non-words (or better, concrete objects/events?) for recognition reaction time. Again in this case, we must reckon with normal context, i.e., coffee as normally stirred, as providing the shortest time. But there should eventually be some equivalence class of liquids which have been primed. As we vary the parameters of the substance, for example, now moving to a thick, batter-like substance, this effect should become more pronounced. Now something like "batter" should be primed more quickly, or "dough," etc. Conversely, as parameters diverge from the coffee stirring event, for example, the diameter of the circular motion grows too large, or the periodicity too different, etc., categorization times to associate words such as "coffee" will increase.

There are "cross-modal" invariants that may be manipulated, crossing between the auditory and the visual for example. Let the event be pouring water into a glass, an event invariantly accompanied by a lawful or invariant increasing rise of pitch.[27] In this case the pitch may be manipulated to actually fall or rise in a manner not coordinate with the invariance law. This, we can propose, should disrupt or lower categorization response time of words, e.g., POUR. This parametric variation cannot be dealt with in current memory models for none hold that the invariance laws defining events are important, nor is there any room in these static models of memory for such laws.

There are many other applications of this redintegrative model to current memory research – many aspects of the findings in Subject Performed Tasks, or imagery effects, or the famous concreteness effect involving the superior memory for concrete words over abstract words, and more. "More" would also include a series of experiments by J. J. Asher on second language learning, where remarkable learning results were obtained using the ecological case – actually coordinating second language learning with performing concrete actions, that is, the subject performed tasks taken in the context of language learning. To Asher, first language learning, on which the learning of a second language should be modeled, takes place in an environment of linguistic commands to which the body/brain becomes attuned – "Johnny, wipe your nose!" or, "Johnny, pick up your spoon!" I have discussed some of this elsewhere.[28] But let us turn to the favorite of the current static memory models – the connectionist networks – and see how these fare in the light of redintegration based upon the invariance structure of events.

Neural Networks versus Redintegration

Neural network models or "connectionist" models have, in essence, been trying to account for redintegration for some time, sometimes in models of paired-associate learning, sometimes in models of semantic learning, though both cases, we shall see, are rather equivalent. The realm of connectionist theoretical effort is vast. The subjects to which the approach has been applied span object recognition, categorization, sequential action learning and disorders thereof, semantic cognition and disorders in semantic memory, consciousness, analogy, and much more. Virtually nothing in the sphere of mind seems out of its reach. In a word, connectionism today is the gorilla on the block; it is the dominant, paradigm-occupying force in the also dominant robotic or machine theory of mind which sees neuroscience as it ally. The ever-expanding breadth of its application is the attempt, with no effort being spared, to show that it is a truly general theory of mind. I am going to describe here two very recent connectionist models, both claiming great power and generality. I want us to see how impoverished connectionism is in the context of real, ecological events. The first model, appearing in the highly prestigious journal, *Behavioral and Brain Sciences*, is by the network theorists, Rogers and McClelland.[29]

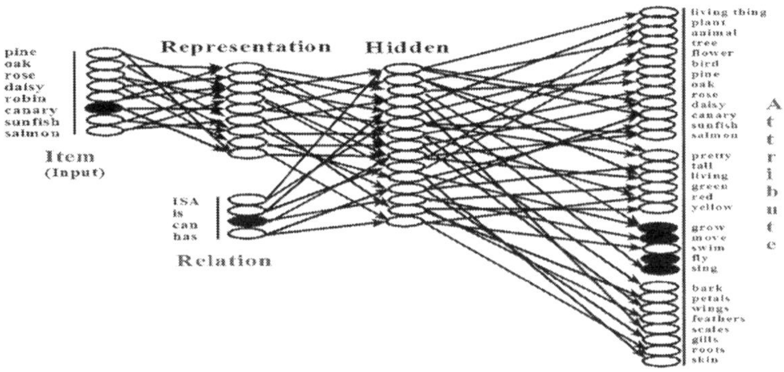

Figure 3.9. Connectionist model of semantic memory. Where connections (arrows) are indicated, every unit connects to every other unit. Adapted from Rogers and McClelland.

A neural network has a set of input units and a set of output units. It also may contain intermediate units in middle layers (known as hidden units). The fundamental goal is to train the output units to respond with a correct pattern of output values for a given input pattern. This is done by adjusting the connection strengths of the various units to each other. Figure 3.9 shows a network of the Rogers and McClelland type that learns "semantics." Thus when the Input unit (on the left) for CANARY is turned on or firing (in black in the figure), as well as the Relation input unit, CAN, the network is trained to respond (or fire) with an appropriate output pattern. In this case, for CANARY CAN, the appropriate responses or output pattern is: GROW, MOVE, FLY, SING.

Adjusting the connection weights of the network takes many "epochs" or cycles of training – giving the network an input, adjusting the connection weights so the output units fire with the correct pattern strength. This is termed "error training." For a given input, such as CANARY CAN, the "error" of each output unit, that is the numerical distance from its correct response value, is measured. The connection weights are then adjusted by a small amount until, ultimately, the network responds with the correct output pattern values, for every input, every time. In essence, this form of network, when the connection weights are all adjusted appropriately, appears to account for semantic learning. It knows what a canary CAN do (grow, move, fly, sing), or what is HAS (wings, feathers), or what it IS (living, yellow), or that it IS A (living thing, bird). SALMON should have its own pattern, for example, that it CAN (swim, grow, move), or HAS (scales, gills), etc.

The fundamental principle operating to achieve the correct output unit states is termed *coherent covariation*. The network is in effect learning statistical probabilities, essentially how things tend to occur (or co-vary) with one another. The fact, for example, that "flying" and "singing" tend to occur consistently with robins while the color "red" may occur with good likelihood for both flowers and robins tends to pull the connection weights apart in the appropriate direction.

Rogers and McClelland argue that this form of network is quite comfortable in the ecological world, that is, it is quite capable of dealing with concrete events. If a child sees a robin sitting on a branch, then sees it fly away, his neural network is being trained. The next time the child encounters the input ROBIN, in the context of a tree branch, he will anticipate that the robin will FLY away. The network is being trained to predict the "sequelae" of various situations, i.e., the events expected to follow in time.

Ecological Events and Neural Nets

While Rogers and McClelland assert that these networks have entered the ecological realm of events, we have seen the actual complexity of the laws describing the structure of even "simple" events, for example, "stirring coffee in a cup, using a spoon." We saw there ratios of cup size to the texture gradient of the table specifying size constancy, tau ratios for modulating the hand, flow fields for specifying form, an inertial tensor defining the "wielding" of the spoon and describing the spoon's resistance to angular acceleration[30], an adiabatic invariant relating the energy of oscillation to the frequency of oscillation[31], and more. Note that this is extremely concrete, *physical* information – forces, masses, weights, motions. It is as concrete, as physically *dynamic,* as the fields and forces of an electric motor. It is this entire informational structure and far more that must be supported, globally, over time, by a neural network – by the resonant feedback among visual, motor, auditory, even prefrontal areas of the brain that deal with plans for action.

Now we can imagine that we have trained the network of figure 3.10 with yet other terms. For example, as input terms, we might have had SPOON, FORK, HAMMER, SAW. As output terms, we could have STIR, CUT, STICK, POUND, TOOL, UTENSIL, etc. In Rogers and McClelland's formulation, I would present SPOON in the context, CAN, and the network is trained, looking at errors in the responses and making the needed connection weight adjustments, to respond with STIR. In the causal "interpretation" of the network's response, after the example of the robin on the branch, the network is predicting the accompanying events we'd expect on seeing a SPOON moving within the cup of coffee, for example, the circular motion of the coffee liquid, the clinking sound, the haptic or muscular feel of the periodic motion, etc.

But what sense is this? This is a remnant of the supremely non-ecological verbal learning tradition, its roots in the semantics-eradication program of Ebbinghaus, which ultimately bifurcated events arbitrarily into components, e.g., SPOON and STIR, then asked: how do we learn these components as a paired-associate pair?In reality, we are perceiving the spoon as an integral part of a stirring event, with all the event's ongoing invariance structure. Where is the "error?" We need no weight adjustments to "link" the event "components" to a proper output pattern. The "components" are already integrally "linked" in the dynamic physics of stirring. For there to be a stirring at all, there must be a liquid, its resistance, an instrument moving the liquid, a force behind it, a spatial constraint on the liquid (such as a cup), a periodic motion, etc., etc. These are not "components" that must be laboriously associated.

A spoon scooping and lifting oatmeal is yet another event with an integral and complex set of forces and auditory/visual patterning. A spoon stirring pancake batter is yet a different complex invariance structure. A spoon digging into and cutting grapefruit is yet another. A spoon balancing on the edge of the coffee cup another. Now we have the set: SPOON CAN: [stir, balance, scoop, cut]. Again we can ask, if I re-present SPOON, which event will it redintegrate? Again, SPOON is roughly equivalent to a static event, a resting spoon. There is little structure to this cue-event; it is underspecified, yet common to all. It is like sending an imprecise reconstructive wave with little coherence through a hologram – we reconstruct a composite image of multiple recorded wave fronts, in this case, spoon-related events. It is the classic interference of McGeoch. The brain's neural network does not require "error-training" to specify this set.[32] This is a completely crazy, Ebbinghaus-like concept of the neural network theorists.

But, as we have seen, we can retrieve specific events. The cue must have the same invariance structure or a sufficient subset. For the coffee stirring, we re-present an abstract rendering of the coffee's radial flow field, or simulate the inertial tensor of the wielding. We are creating, globally throughout the brain, a more constrained, more coherent "reconstructive wave." To reconstruct the batter-stirring event as opposed to the coffee-stirring, we must constrain our wielding cue differently to capture the larger amplitude of the stirring motion and/or the greater resistance of the batter.

The "coherent covariation" that these networks detect is essentially a low order form of invariance, what I shall term a *syntactic* invariance. Though we will discuss syntax and syntax rules more fully in Chapter VI, syntactic invariance is of a very low order; as low an order as facts like the nonsense syllable QEZ always occurs with PUJ, or WUX with GEK. This is but a simple spatial conjunction of objects. Object X is always found by Y. And this is the very definition of syntax: *laws for the legal concatenation or juxtaposition of objects.*[33] It is a purely *spatial* operation – in the sense of the classical abstract space. The lists of nonsense syllables invented by Ebbinghaus were nothing but low order syntax rules. To the neural network, its inputs and outputs are just as meaningless – pure syntax rules. This level or form of invariance is utterly insufficient to capture the dynamic transformations, forces, fields and invariance involved in time-extended events like coffee stirring. This is where the *semantics* resides.

The connectionist will wish to invoke *recurrent* networks as the answer to handling dynamic events with invariance defined over time. There are many complex variants of recurrent networks, but taking recurrence at its most elemental beginning, it is clear that if we begin with our standard network (as in Figure 3.9) with its input layer, output layer and hidden (the middle) layers, and simply add a "context" layer of units which holds the previous state (or weight) values of the hidden units, this does not change the basic pattern – we are still learning syntax rules and in a discrete state model of time. Recurrence, in and of itself, has nothing that intrinsically allows it to represent the dynamics of these events and the forms of invariance existing over their time-extension that we have discussed. If perchance it does, if recurrent nets can support wielding, stirring, and rotating cubes, this needs to be explicated explicitly relative to the ecological laws of events.

I will reemphasize the previous sentence. Ecological psychology is a *science* doing the appropriate work of a science. It is discovering invariance laws. Taking a function of mind such as semantic learning and creating a mechanistic model that approximates this function – and, just as with Newell and Simon's GPS, an approximation is all that connectionist models ever achieve – is not all there is to science. It is incumbent upon connectionism to show how its mechanistic networks meet science, particularly the science of ecological psychology. Connectionism must show how its networks support the invariance laws defined over the concrete events of the ecological world. I, for one, hold that they cannot. The burden of proof is on connectionism.

Neural Networks versus the Turing Test

It is a simple fact that we have no problem understanding, "A SPOON can CATAPULT (a pea)," for the spoon, we know, can be inserted in, and support the forces/invariance structure of, catapulting. Does this mean that the network would have to be trained, specifically, with CATAPULT as an output unit/term? Absolutely, for according to the Rogers and McClelland model, without the appropriate connection weights being trained up, the network would have no "knowledge" about this fact. Does this mean that the network has to be trained for events like, "A spoon can hammer," or "A spoon can mash," or "A spoon can make music," etc? Yes, it does. What this means is that the network would have to be trained for every possible, unforeseen event in which a spoon might participate, or in which we might decide a spoon could participate.

Here we have entered the realm of a critique by the computer scientist, Robert French, on the possibility of computers ever passing the Turing Test. The "Turing Test" is a well known procedure devised by Alan Turing, one of the pioneers of computing. Its purpose is to test present and future computers (and their software) to determine if the computer has arrived at the point where it is indistinguishable from a human. The procedure is for the tester to ask written questions of a computer placed behind a screen. The answers are returned in written form. Many think it inevitable that a computer will someday pass this test. For Ray Kurzweil, this should be well before 2045, the year when robots eclipse us.

French gave a devastating critique of this possibility.[34] He argued that no computer will ever pass a test where its answers were compared to humans for questions such as:

- "Rate purses as weapons,"
- "Rate jackets as blankets,"
- "Rate knives as spoons,"
- "Rate banana splits as medicine."

Humans easily can determine whether a purse can make a weapon (for self defense) or a jacket can be a blanket. Some of these depend on the transformation. Knives can stir coffee just fine, but are lousy for eating soup. But computers would need to store the *experience* of the human being. For this, symbols, syntax rules and data structures will not do. Nor will neural networks. The problem equally holds for evaluations of statements such as:

- "A credit card is like a catapult,"
- "A credit card is like a fan."[35]

The list is endless. Says French, "...no a priori property list for 'credit card,' short of all of our life experience could accommodate all possible utterances of the form, 'A credit card is like X'."[36] But what is experience? What is it other than perceived events, ongoing over time, which have an invariance structure? Without the ability to support the invariance structure of events, the neural network models have no chance of "storing" the requisite experience. Lacking this, as French too noted, ultimately these networks, yes, the very networks of Rogers and McClelland, would have to program (or train the connection weights) for *all possible pairs* of objects and actions – and even this massive syntactic rule pairing would still be woefully insufficient. Re this critique, effectively an impossibility proof for connectionist networks, the connectionist theorists' response has been: Insert head in sand.

What is happening, then, when we rate "a knife as a spoon?" It can be described as this: *it is the projection of the transformational dynamics of an invariance structure upon a possible component.* We place the knife within a coffee-stirring event and see if it can "hold up," that is, see if has the requisite structural properties to move the coffee medium. So too, when we rate credit cards as spoons (under a stirring transformation) or as fans (under breeze generation), or spoons as catapults (under a catapulting transformation). The list is endless because the card or the spoon is fair game for an infinity of transformations under which new structural invariants can emerge. More precisely, each is fair game for insertion into an infinity of invariance structures.

The Analogy Defines the Features

Note that these statements – a credit card is like a catapult, a credit card is like a fan, a spoon is like a catapult – are *analogies*. To account for analogy, one cannot begin with a preset list of features. The features emerge under the transformation. In effect, *the analogy defines the features.*

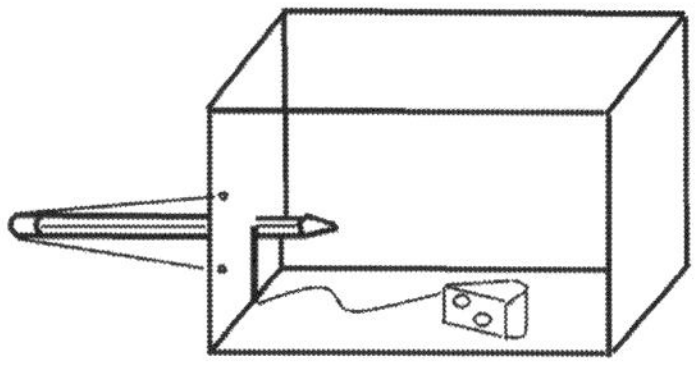

Figure 3.10. The Crossbow Mousetrap

Consider being asked to design a mousetrap. We are provided with several components: a piece of cheese, rubber bands, a 12" cubical box, pencils, a razor blade, toothpicks (the strong kind), a rubber eraser, string, tacks. We might attempt to list all the "features" of each object. But what actually happens in thought? Our aim may be killing the creature, but there is no useable content in a purely abstract "killing." Killing *is* an invariance across concrete forms of killing. So perhaps

I contemplate crossbow shooting (Figure 3.10). This places the potential components within a dynamic transformational structure. The stretch-ability and force of the rubber bands emerges, the sharpness and straightness of the pencils, the "anchoring" potential of the side of the box to which I will tack the rubber bands, etc. The toothpick holds the pencil back; the string is attached to the cheese and toothpick. These features become a (newly created) "object language" I could provide a solution program. Or, contemplating beheading by axe, the length and requisite strength of the pencil emerges. I can groove the pencil and wedge the razorblade in to make an axe. The "container" property of the box corner emerges, as I can prop the raised pencil-axe in the corner, a toothpick will prop it up, a rubber band tied to the pencil and tacked to the "anchoring" feature of the floor will provide downward force, etc.

These "features" or "properties" of the objects dynamically emerge as a function of the transformations placed upon them via the invariance structure and the constraints naturally specified by the proposed structure, e.g., the crossbow requires anchoring points for the bowstring – the rigidity of the box can provide these. They cannot be all pre-set. New ones will always emerge. This is the problem with any approach that holds that the features define or determine the analogy.

Life Defines the Features

The concept that analogy defines the features is profound. It is pivotal to the nature of mind and thought, completely determining the nature of the "device" that we actually are. The concept is not originally mine, though it may have first appeared in my 1976 thesis, I don't really know. When I arrived at the Minnesota graduate school in 1970, fresh from the Marines, there was a young graduate student in psychology there named Jim Reeves. Jim was not too sure of me. He was somewhat of a flower child with long hair, a beard, and an anti-war philosophy. He had a degree in physics and was brilliant. It took him a couple of years to come to the opinion that I was not a complete idiot

Jim and I took a canoe trip one September in Quetico Park in Canada, just above the Minnesota border. We were out nine days, and traveled over a hundred miles. At the northernmost apex of the trip, probably forty miles above the border, we camped on an island in a small lake named for the poet, Keats. Two wide, powerful waterfalls poured into the lake on its eastern side, separated by perhaps a quarter mile, and about a quarter mile distant from the island. The whole lake moved by the island and onward to the powerful Splitfrock falls at its western end. Sitting that night by a fire on the rocky shore of the island facing the two falls, with the moon hanging in the sky over the mists rising from the tumbling flows, Jim and I talked about the theory of mind, language, Shaw's notion of "coalitions" as the form of organization of the brain. In this discussion, Jim, in his slow, considered style, stated out of nowhere that he was coming to the conclusion that, "it is the analogy that defines the features." I did not understand the implications at the time, but what Jim had seen was radical, far beyond the grasp of the computer modelers of mind – to this day.

But Jim did not live to develop his vision. His family had a history of schizophrenia. One night a few months later, Edna Green, a mutual friend and grad student, phoned me to come over quickly. Jim was in a strange, near-ecstatic state. It seemed to me a near-mystical state. He spoke of being in a new state of consciousness, one towards which many of us were moving. When I left, he seemed ok, but this state degenerated later, early in the morning. He began seeing Nixon flying around the sky. He was taken to the hospital and was eventually put long term on a drug that suppressed the episodes, but made it impossible for him to concentrate, destroying the life of the mind in which he had his being. He became so depressed over this, he slashed his wrists a few months later. We lost the unique exposition he would have given us.

AI and the Problem of Analogy

Analogy is the fundamental basis of thought and language. The just previous sentence relies on analogy; it is invoking the concept of a basement or foundation underlying and supporting a structure, in this case the structure of thought. Metaphor, which is simply another form of analogy, is the cognitive basis of language. Language is littered with "dead" metaphors that are simply not recognized as such, for example, "He grasped the concept," or "He jumped right on the task." If you cannot support analogy, you have no theory of intelligence. You cannot support thought. You have no theory of cognition.

AI comes in two major flavors. One, neural networks, we have just seen. The other can be termed the "symbolic manipulation" paradigm. This latter was the first to be developed, starting roughly in the 1960's. This too we have seen. The GPS program of Newell and Simon (Chapter I) was an example. Connectionist models are considered more near the biological, and have the advantage of a natural learning process through their weight adjustments. Both approaches continue to vie for position; in some camps they are seen as complements, though no one knows how the symbolic manipulation model can use the output of a neural network, i.e., how there can actually be a complementary relationship. It doesn't actually matter. Neither can support analogy. Both end at the same place.

We have seen (some of) the problems of the neural networks in this regards. The symbolic programming method has proffered several models for analogy making, the most famous of these being the Structure Mapping Engine (SME).[37] To SME, as in all AI, the *features* define the analogy. Thus SME treats analogy as a mapping of structural relations relative to pre-defined features. The solar system, for example, and the Rutherford atom both have specific features and their relationships described in predicate calculus form, e.g., Attracts (sun, planet), Attracts (nucleus, electron), Mass (sun), Charge (nucleus), etc. Douglas Hofstadter (author of *Gödel, Escher, Bach*), with his former students, yup, Robert French and David Chalmers, level a heavy critique upon this approach, noting the helplessness of SME without this precise setup of features and relations beforehand, and with this setup given, the purely syntactical, nearly "can't miss" algorithmic or proof procedure that follows. The resultant discovery of analogy is, to quote these critics, a "hollow victory."[38]

BACON, as another example, attacks Kepler's problem of discovering the law of planetary orbits. It quickly solves the problem with a precise, tabular representation of the solar system showing a primary body (Sun), a satellite body (planet), a time T the two objects are observed, and two dependent variables – the distance D between primary and satellite, and the angle A found by using the fixed star and the planet as the endpoints and the primary body (Sun) as the pivot point.[39] Kepler, French notes, took 13 years to sift through the data and flawed concepts of the solar system to find the relevant features. Yet SME, he notes, uses an entirely different representation of the solar system, exactly suited to its programmatic purpose, to find an analogy to the Rutherford atom.

The features on which the analogy is based cannot be preset, pre-defined. As noted, it is the analogy that defines the features. Analogy is a transformation. This is to say that it is a process that occurs over an indivisible, non-differentiable flow of time. It is supported only over concrete experience or the remembrance thereof, i.e., it is carried only over the transforming images or the figural mode. AI – based in an abstract time and without a theory of perception – can support neither of these requirements for analogy. It has, therefore, not a prayer as a theory of cognition.

An AI theorist, Eric Dietrich, in an essay (2001) wherein he arrived at the principle that it is the analogy that is defining the features, discussed what he termed "analogical reminding."[40] Walking in an alley, he spots a configuration of garbage cans. Instantly it becomes "Garbagehenge." In essence, the static structure of the garbage cans has driven recall of past experiences of Stonehenge, whether visits or pictures; the lowly garbage cans are now exalted in what we can call an "analogic" event. In this little example, Dietrich was struck by the fact that at the basis of analogy is an apparently instantaneous operation of memory retrieval that, of itself, is inherently analogic. As a computer theorist, Dietrich had no idea how this could be implemented. We have seen that this operation is redintegration. The lowly garbage cans were simply a "static" event. We have been dealing with events that are far more dynamic – continuous transformations or change in the external field. Connectionism begs the description of this change. It starts after it has crystallized and frozen this change into objects and "properties." But it is real, concrete, changing events that drive the brain. Events have a structure and the neural architecture must respond to this structure. It is the invariance laws of events that drive the fundamental memory retrieval operation supporting analogy.

DORA: An Explicit Connectionist Model of Analogy

Though the Rogers and McClelland model is in effect a model of analogy, a bad one, it is not explicitly billed as one. We are going to look at one other connectionist model, one directed very explicitly to analogy. The connectionists know they must account for this, and this model, appearing in the prestigious journal *Psychological Review* is both typical and perhaps the height of their efforts. My intent here is rather ruthless, for I wish to erase all doubt that the neural network or connectionist paradigm has the proverbial snowball's chance of accounting for this all-important function of mind.

Doumas, Hummel and Sandhofer present a model (DORA, or the Discovery of Relations by Analogy) for the learning of relational concepts.[41] The model begins with what the authors term an "holistic representation of objects." A "cup" would have an associated set of features, represented by semantic units (Figure 3.11a). A "proposition" in this framework – simply a sentence like, "A cup is bigger than a thimble" – is represented in the form: *bigger*(cup, thimble). Propositions are represented in four layers (Figure 3.11b), with the just-noted bottom layer of semantic units encoding features. These "features" the authors term "invariants" (very questionably we shall see) include things such as "visual invariants" (round, square), relational invariants (more, less, same), dimensional invariants (size-3, size-5, color-red), complex perceptual/cognitive properties (furry, barks), category information (apple, dog). (In this scheme, "size-5" is a larger size of something on the "size" dimension than a "size-3.")

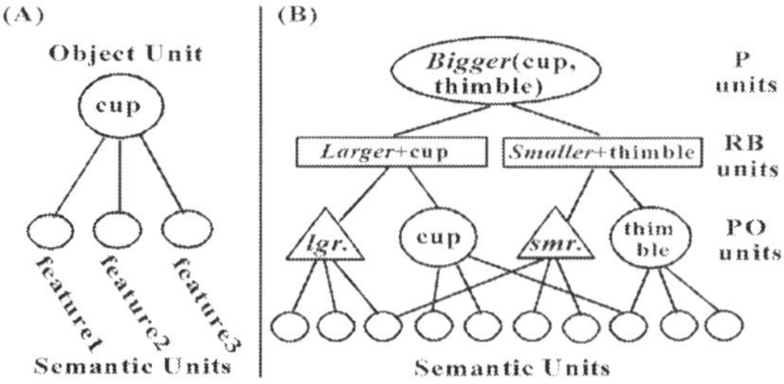

Figure 3.11. The structure of propositions in DORA.[42]

The units of the second layer – predicate and object units (PO) – code for individual predicates (relational roles or attributes) and objects. A symbol for the object *cup* in the proposition *bigger* (cup, thimble) is connected to a set of semantics corresponding to the features of *cup* (round, handle, shiny) while the symbol for the relational role *larger* is connected to a set of semantic units corresponding to its (the relational role's) features. The next layer, the role-binding (RB) units, encodes the bindings of relational roles to their "fillers." In our sample proposition, *bigger*(cup, thimble), cup and thimble are the "fillers" in the relation *bigger than*. Thus, in the proposition *bigger* (cup, thimble) there is an RB unit representing the binding *larger*(cup) and an RB unit for *smaller*(thimble). The final layer is comprised of a set of proposition (P) units which binds RBs in multi-place relations, where our *bigger*(cup, thimble) is such a multi-place (here a two-place) relation.

This scheme or system, the connectionist theorists like to say, is *compositional*. We are "composing" the multi-place relation *bigger*(cup, thimble) from the lower level elements. Compositionality is considered a big thing in cognitive science theory ever since a prominent theorist, Jerry Fodor, declared it to be a critical aspect of cognition and demanded it as a capability in neural nets (it is just a given in the symbolic manipulation model, i.e., given away for free, just like "features.") We'll visit the true origin of compositionality for a bit in Chapter VI.

We can neglect the model's process for the formation of these connections. What I wish to note immediately is that the authors acknowledge that while the model is developed almost solely in the context of dimensional relations such as *larger* or *smaller* (where the model will employ units for comparisons like "size-3" versus "size-6") they insist that the model will equally hold for the relations involved in *events*. In their example, they use the event of *chasing*. A relation such as *chasing,* they argue, differs from one such as *bigger* only in the semantic features composing it. When a child chases a dog or a butterfly, he can compare the experiences and "predicate what it feels like" to be a chaser. The semantic features attached to the feeling are likely to include those of motion, running, excitement, a sense of pursuit, and the chaser role thus gains semantic content. The same will hold for experiences (and the role) of being chased, and thus the child has the opportunity to learn, via a key operation for the model, namely the operation of *comparing* these experiences/events, that *chase* is a two-role relation in which both objects are in motion, both experience excitement, but one object wishes to catch while the other wishes to escape.

The model, the authors note, does not speak to where the semantic invariants underlying *chases* come from, but it does speak to the question of how they eventually become predicates that can take arguments and can therefore support complex relational reasoning. Their claim is that human cognitive architecture, starting with the invariants or regularities in its environment, isolates these invariants and composes them into relational structures (like we have seen above), and that DORA's learning mechanisms can convert these into explicit predicates and relations that make reasoning possible.

This claim is the critical issue. The claim, we saw, can equally be found, in only slightly less explicit form, in the extensive connectionist model of semantic cognition of Rogers and McClelland. Doumas et al. state explicitly what has been an implicit assumption of connectionism, namely this: a connectionist cognitive/memory architecture can simply inherit the results of perceptual processing – the "semantic invariants" – and begin from there. This is not the case. An "invariant" has been trivialized here. It has become merely a static feature connectionist theory presumes it can use as an abstract object in a rule system. This is the fundamental "simplifying assumption" of connectionism, namely, that the world and its events can parsed into a set of static properties, features, "invariants" that now become food/symbols for a connectionist rule/relating system. We have seen that the fundamental operation of memory, redintegration, is driven by the invariance laws defined over time-flowing ecological events – invariants that do not exist in an "instant." These events cannot be reduced to static properties, features, or "invariants" inherited from perception as connectionist models use this term. The very processes of perception – concrete, physical, dynamic, flowing in time and unusable by connectionism – are already the basis for, and operating in, redintegrative memory retrieval, long before connectionist networks even engage in the subject after their simplifying assumption.

Event "Geons" and the Real Origin of Event "Features"

The *spoon-stirring-coffee* event is a dynamic transformation of a portion of the ecological world. Something would be required to accomplish the work of partitioning it into the semantic features with which connectionism begins. Rogers and McClelland imply it has been partitioned into STIR and SPOON, with SPOON partitioned into the features it HAS and IS. But STIR is a higher order concept covering a very complex transformation. Under this, the spoon is "creating a force," it is "pushing the liquid," the liquid "is moving in a radial, complex pattern of motion," the spoon is "clinking" against the side of the cup, it is "constrained" to a certain radius, it has a "periodicity," it is "rigid," it has a certain "flatness," but has a "cup-like quality," it is "shiny." All these are complex patterns, for the visual motion as complex as a flow field, or for the hand's motion as complex as the "wielding" of the spoon where there is an inertial tensor (a set of forces) and adiabatic (frequency/energy) invariant. The above is an arbitrary set. None are truly separate, the *event* does not happen without the spoon, the forces, the liquid, the container.

Meanwhile, across the table, big sister is using the spoon to scoop cereal. Again, another complex transformation. The spoon is applying different forces, a different motion, a different visual transformation; the Rice Krispies are parting, piling into the spoon, falling off. Or little brother is using the spoon in ladling soup – another complex transformation. Yes, it makes sense to ask, given a SPOON, what can it do? But we know it makes no sense to be attempting, via weight adjustments, to *re-link* spoon to stir, or spoon to scoop, or spoon to ladle. These events (spoon stirring, spoon ladling) are already inseparably whole events. What does make sense is to ask how SPOON, *as a cue*, can redintegrate a set of *events* in which it participated. But this is an entirely different operation, as we shall discuss.

Now we imagine a set of stirring events – a spoon stirring cereal, a spoon stirring soup, a spatula stirring cake batter, a board stirring cement (Figure 3.12). Considering just the general visual form – the radial flow field – there will be an invariant form across all these events. But to obtain it, there is first an important and disturbing implication for memory. As a position in memory theory, known as *exemplar* theory, insists upon, *all* our experiences of these stirring events must be stored.[43] The abstract invariant, *stirring-motion*, does not exist unless the separate events also exist across which the invariance

Figure 3.12. A "stack" of coffee stirring events.

can be defined. Given this stack of events, we could obtain an abstract feature, call it "swirl" – in truth, merely a higher order invariance across the radial flow patterns of all these events. This we could "attach" or link to an abstract STIRRED relation-unit (or STIRRER? Both?), though the notion of a "link" between it and sub-units becomes clearly a bad misrepresentation of the reality here – it is already inextricably part of all the events over which it emerges as an invariance. So also we could "attach" an abstract "wielding," an abstract "clinking," an abstract "constrained-to-a-circle" motion of an abstract instrument, an abstract "resistance to the motion," an abstract "immobility of the container," etc., etc. And how could this operation, defining an invariance across the "stack" of events, happen?

We require an operation that takes an entire stack of fully specified (multi-modal, time-extended) events, much like the computer scientist, Gelernter envisioned, collapsing them such that the invariants stand out, the variants becoming noise.[44] As we have seen, the classic, concrete embodiment of such an operation is exemplified by the recording of a succession of n wave fronts on a hologram. In Figure 1.4 we used different objects – a pyramid/ball, a cup. Now we imagine using simply different cups, where each cup is of a different form. With each object wave reflected from a given cup, we pair a unique reference wave of frequency, f_i, creating a unique interference pattern for each recorded object-wave/reference wave pair. By then modulating a reconstructive wave passing through the hologram to each successive frequency, f_i, we reconstruct the image of each cup uniquely. But if we employ a less than coherent reconstructive wave, a composite of f_i thru f_n, we will reconstruct a fuzzy image. The invariants across all the cups will yet stand out, the variants being confused. In this operation, we did not begin with separate features of the cup – the cups were a whole – and only through the operation did separate "features" stand out.

These higher order invariants – in essence, classes – emerging via our reconstructive wave operation, are the "invariants" DORA intuitively assumes exist, and which it can use. But these exist only over the complex invariance structures and dynamic time-extended patterns we have discussed. It is not just cups, it is time-extended events like stirring. The invariants' emergence is dependent, in this scenario, *on a form of redintegrative memory operation*, retrieving sets of whole events and their structures, that is radically different from the memory operations DORA envisions (and which we will see it subsequently employ). Even with such an operation, we would yet question how or why a separation operation now occurs to carve the invariants into independent "feature" units that are now linked to a "relation" in a loose conglomerate.

Curiously, Doumas et al. approvingly invoke the JIM or "geon" model (see Chapter II, Figure 2.2) as a prime example of one of connectionism's "triumphs," in this case in the realm of object recognition. We have seen that this is absurd; Shaw and McIntyre's wobbly cube destroys this static model. Further we noted that the geon turned out to be a handy way to hold the features of the cube together as a whole, else the separately stored features could be put back together in all kinds of ways, none looking like the original cube. The "geon" here was the handy picture on the top of the puzzle box that gives us a clue on how to fit the puzzle pieces together. Yet we have the same problem for the "events" of connectionism in general. Even if we had this set of complex features linked to the relation, we are in the same shape as in the decomposition of the cube and cone into features. Now we need "event geons," and particularly here, a *stirring geon*, for in the decomposition of the stirring event (both arbitrary and with no definable limit to the number of "features") into static feature units attached to an abstract "relation" unit, we have lost the global picture of the transformation.

If you have lost the global picture of the transformation, you have lost the ability to form analogies. You can no longer place a pencil in a coffee stirring transformation or stirring event and see if it will work, that is, see if the requisite "features" will emerge. Nor can you place a spoon in a catapulting event. You have no events. You have no analogies.

The DORA Comparator: "Features" = Whole Events

DORA employs a comparison process across what it calls "events," or better, partial fragments of events, but it starts this higher up its hierarchical structure (of Figure 3.11). As it assumes there is a feature-fragmentation accomplished (somehow) by perception, let us consider this further for a moment. At this level, DORA envisions comparing object units, finding common features and creating single place predicates. Thus we might bring *spoon* and *spatula* into memory, and find a common "feature," namely, "stirrer." This stimulates the birth of the explicit property, "stirrer," and this can be bound to each object, such that we have *stirrer*(spoon), or *stirrer*(spatula). *Every* feature in common between the two objects, according to DORA, is also linked to this relation (*stirrer*).

This is the first problem: If "stirrer" is a feature of these two objects, how can this "feature" possibly be other than a full stirring event with its invariance structure? Just what is a "partial" stirring event such that we only have a "stirrer" as a feature? Now, to find these two "features" of the two objects the "same," we must be implicitly comparing the two *events*. But the two events – spatula stirring of a cake batter and coffee stirring – have similar but different radial flow fields, different values of wielding, periodicities, acoustics, etc. How does the model (with only its abstract symbols and tokens) determine these two "features" are the "same?" How does it does it achieve this without an actual physical comparison?

The second problem: Figure 3.12 shows the dilemma, for there I attached features to spoon such as wielding-1 (i.e., some arbitrary "value" for wielding the spoon), circling, rigid, etc. Are these part of the object, or part of the event in which the object participated? In reality, they are part of *stirring*. If they are part of the spoon, how are they the "same" as the spatula's?

The third problem: In DORA, *all* features the two objects share in common are attached to the newly formed *stirrer* relation. The spoon occasionally sits on a surface, so does the spatula. Both are occasionally washed. DORA contains a "pruning" process to rid predicates of such extraneous features, but unless two objects are compared, where one object stirs but is never washed nor rests on a surface, these features could not be pruned and would remain with *stirrer* forever. This is nonsensical. How could these features (resting, being washed) possibly have been part of a stirring event (relation) in the first place?

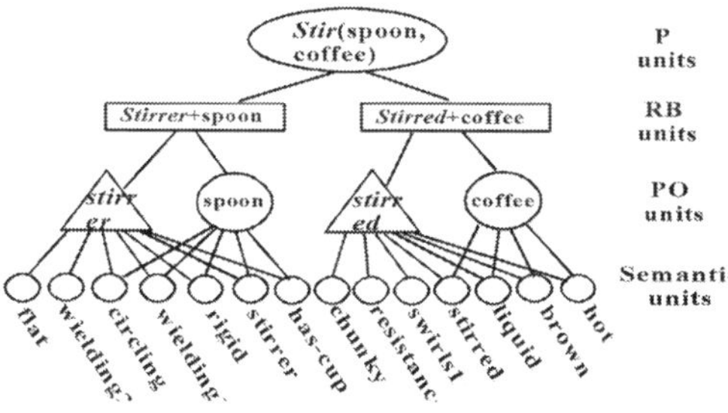

Figure 3.13. A DORA-like proposition structure for *stirring*

The origin of these dilemmas is the fundamental "simplifying assumption" of connectionism whereby the ecological world is viewed as objects and properties, with properties attached to objects. But objects are simply special cases of an event. In turn, an object-event is always itself nested in some event – the spoon at rest is an event, or the spoon is stirring, it is scooping, it is cutting cheese, it is catapulting a pea. Its "properties" emerge over or during the specific event. It is events that are primary. And events have a structure over time that is resolutely ignored only with the consequences above. The cheese-cutting spoon, the soup-stirring spoon, the pea-catapulting spoon – all show different properties that are not "directly apparent" in the resting spoon. "Properties" of objects are always functions of events.

DORA the Brittle

I can take the analysis of DORA relative to actual, ecological events no further, for other than the brief comments on *chasing*, the model's mechanism for actually engaging concrete events is in fact non-existent. There is no algorithmic engine to create the proposition *stirring*(spoon, coffee) from the single place predicates such as *stirring*(spoon) (which themselves have the strange origins noted). The current engine is a comparator that operates on propositions which have a dimensional value (for example, dimensions such as size, height, length, weight). If DORA "thinks" about a DOG of size-6 and a CAT of size-4, the comparator, detecting the dimensional value, links a "more" relation to the size-6 (or "*more*+size-6") related to the DOG and a "*less*+size-4" for the CAT. If this pattern reminds DORA of a previous comparison of the same type between a BEAR (more+size-9) and a FOX (less+size-5), a further operation now compares the CAT and DOG units to the similar setup for the BEAR and FOX, eventually spawning a new unit, BIGGER, bound to BEAR and FOX or *Bigger*(BEAR, FOX).

Given the complexities of the subject of analogy with the concrete dynamics of the events involved, this is an extremely low-horsepower analytical engine, something like opening the hood of your Mercedes and discovering forty-two chipmunks winding a rubber band. Obviously, a quite different engine is needed to create an ecological "predicate" such as *stir*(spoon, coffee). There are no handy, nicely programmer-provided values such as "size-6" or "size-4" to rely upon.

I will submit that any system proposed will have, lurking in the background, the implicit knowledge of the dynamic structure of each event (e.g., *stirring*) it is dealing with. Without such event structure constraints, what would prevent DORA from thinking about *stirrer*(spoon) and *chased*(Mary), and being reminded of a previous comparison, *stirrer*(spatula) and *chased*(Joe), proceeding to derive *stirring*(spatula, Joe), i.e., the spatula stirs Joe?

The proposition, *stirring*(spatula, Joe), is the essence of a syntactic "failure of reference." As a sentence, it takes its place with other sentences that are syntactically correct but seem to have no semantic justification:

1. The leaf attacked the building
2. The shadows are waterproof.
3. The spatula stirred Joe.
4. The building smoked the leaf.

Katz and Fodor, early in the game, tried to solve this problem by "semantic markers" assigned to each lexical item in what is termed the "deep structure" of a sentence.[45] These "markers" were simply syntactic rules trying to represent physical constraints – rules attempting to do the work of the invariance structure. The "leaf" in (1) would thus receive a marker denoting it as *inanimate* among other things, while "attack" would receive a marker requiring its use with an *animate* object. Having incompatible semantic markers, such a system brands the sentence as meaningless. "Stirring" would have been tagged with a marker requiring its object to be, say, liquid. Joe, having no such marker, would have thus been seen as illegal in (3) and the string also branded as meaningless. Unfortunately such sentences can appear very meaningful. Sentence (2) would also have incompatible markers, yet is perfectly interpreted as meaning that we can throw as much water on shadows as we like and they will be unharmed. As for (3), we *can* easily make sense of this sentence, "The bad architecture of the software system is like a spatula, stirring Joe, the programmer, into an anxious mess." Such facts quickly lead to "rules for relaxing the rules," but the rule system quickly ends in anarchy, being so flexible that it is useless as an explanatory device.

What we have been exploring here is the greatest unsolved problem of AI. It is termed the problem of *commonsense knowledge.* Analogy is simply an aspect of this general problem. The ability to see that purses can be weapons or spoons can be catapults is part and parcel of this general problem. Its source is in AI's inability to deal with the invariance structure of events, itself a problem of perception and time. We will see this in another light (Chapter VI) when we visit the support invoked from AI by apologists for the theory of evolution.

DORA: Fully Mired in French's Impossibility

In the DORA model, the creation of BIGGER rested on the setup of the dimensional features, e.g., size-5, feeding its comparator. Ignoring for the moment that DORA's comparator is not even close to something that can handle actual, ecological events, let us suppose we have formed single place predicates (SPs) such as *stirred*(coffee), *stirrer*(spoon), and *stirred*(paint), *stirrer*(paint-stick). According to the model, a pair of single place predicates enters working memory, in this case *stirred*(coffee) and *stirrer*(spoon). These are "mapped" as a unit onto other SPs, in this case *stirred*(paint) and *stirrer*(paint-stick). This mapping serves as a signal to link the SPs into a larger predicate structure, thus *stir*(spoon, coffee) and/or *stir*(paint-stick, paint).

This is simply a syntactic mapping. It is simply a proof procedure carried out in a fixed "object language" or categorization of the world. It has nothing to do with analogy. It is based on the fact that we attached "stirrer" as a feature to spoon, and "stirred" as a feature to coffee. Given the precise setup of the SPs, the mapping can occur via an algorithm. Without this precise setup, the process is helpless. The network has no ability to recognize the validity of, or create, multi-place predicates such as *stir*(knife, coffee) or *catapult*(spoon, pea) without this setup. Again, it is another example of the validity of French's critique. There is nothing in the network, unless it has been specifically trained, and the "features" specifically set up, that would support these relations. The knowledge which supports these resides in the invariance structure of events, and these structures are nowhere near representation in this model. The real problem, we saw, is the ability to *create* object languages.

We have already noted the helplessness of these kinds of analogy models without this "precise setup," for example, SME's treatment of the solar system and the Rutherford atom with the features of each nicely predefined, and with this setup provided, the purely syntactical, nearly "can't miss" algorithmic procedure that follows, and, as noted, the resultant "hollow victory." In essence, connectionism has moved into precisely the same critical, problematic area and taken on the same ills as the symbolic manipulation paradigm. This is to say that it is equally helpless before the problem of commonsense knowledge.

Experience has taught me that connectionist theorists, invariably failing to grasp the general point here on the existence and nature of invariance structures defined over dynamic events and the necessity to actually engage invariance laws, will simply point to some other connectionist model undiscussed here, this other model being the "answer." Indeed DORA is but one of several implementations of connectionist architectures which aim to implement the construction of analogy. One that was pointed out to me as purportedly far superior to DORA is the Vector Symbolic Architecture. This is an effort to implement a more biologically plausible, more efficient architecture without using the troublesome backpropagation (the manual, error/weight adjustment) method we saw in the Rogers and McClelland network.[46] Nevertheless, the fundamental view of analogy is exactly the same across this and all these models, namely, analogy depends on a systematic substitution of components of compositional structures. The systematicity and compositionality are considered the outcome of two operations, namely, binding and bundling, where binding ties fillers (Spoon, Coffee) with roles (Stirrer, Stirred), and where bundling combines these role/filler bindings to produce larger structures, e.g., *stirred*(spoon, coffee). Therefore the very elements of the composition, i.e., the features of objects, must be predefined, the values set, the relations fixed – all for the sake of allowing a syntactic process or proof procedure to unroll.

There is another famous problem in AI, termed the "frame problem." It is but a variant of this wide-ranging problem of commonsense knowledge. Put generally, the frame problem asks how a robot determines that a change in his environment is an expected change or not. As a robot stirs the coffee, let us imagine that the coffee cup bulges in and out, or floats above the table, or the coffee makes a "snap, crackle, pop" sound, or the liquid provides enormous resistance like cement, or the surface spouts small geysers. Are these expected aspects of the event? Must the robot be checking, instant by instant, its database of "features" of the world to see if these are expected features of his world? To be constantly performing such a massive database check is extremely expensive (read impossible) computationally. The solution to this dilemma remains to be found. We will visit the frame problem in Chapter VI. There is a better way.

Failure to Net Quality

Consider the concept of "mellow." The word has manifold meanings: we can talk of a wine being mellowed with age, a dimension of the word we apply to taste. We speak of a violin being mellow or of a song being mellow, a dimension applying to sound as well as mood. We speak of the interior of a house or room being mellow, referring to the visual. We can say "mellow" of a soil. The concept of "mellow" expresses a very abstract qualitative invariance defined *across* many modalities. At the same time *within* each of these dimensions it is a quality that emerges only over *time*, within the experience of a being dynamically flowing over time. "Mellowness" does not exist in the instantaneous "instant." This quality can only become experience for a being for whom each "state" is the sum and reflection of the preceding "states," as a note in a melody is the reflection of all those preceding it, a being whose "states" in fact permeate and interpenetrate one another. If we take this to heart, we should say that the meaning of the word "mellow" is an invariant defined within and across modalities and over time. It is not a homogeneously represented invariant, nor can it exist in space, when space is defined as the abstract, three-dimensional, instantaneous cross-section of time.

Yet, this instantaneous, homogeneous representation is exactly what AI, in either of its flavors, would propose. The data structure supporting a symbolic program would consist of "nodes" with abstract links to "mellow," as for example in Figure 3.14. We can easily imagine the network of Rogers and McClelland being extended to incorporate "mellow." We would see things like VIOLIN IS {mellow], ROOM IS [mellow], or SOIL IS [mellow]. That these syntactic structures have nothing to do with the meaning of mellow is obvious.

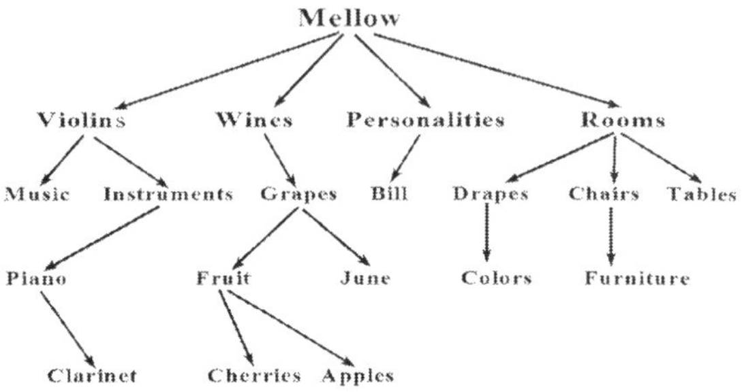

Figure 3.14. A (rough) data structure of nodes with links or pointers to various concepts.

Rogers and McClelland presumably believe that VIOLIN and MELLOW are somehow related to other parts of the brain's neural net as it processes an event involving a mellow violin. But then we are back to the question, where is the "error." Why would the net need training to adjust its weights to "associate" violin and mellow? The mellow quality of the violin is an intrinsic aspect of the event, just as we said in regards to all the aspects of the coffee stirring. Again, these "components" of an event don't need to be "associated." Further, there are untold numbers of these components in an event – nameless invariants – that have no "symbol" but whose existence can suddenly become quite obvious. I say, for example, "As he stirred, the coffee went snap, crackle and pop." This is obviously a violation of an invariance law in coffee stirring. It happens to be the source of humor; one can imagine this scenario causing laughs in an animated cartoon. But what is the "node" or symbol for the nameless invariance being violated here that we would link something to in our neural net? And again, this network cannot possibly account for the emergence of analogy, where suddenly, for example, a culture, as in the 1960's, decides it is cool to apply this quality of mellowness to a *personality*, as in "Joe is mellow."

Is Everything Stored?

Now we must ask: Is *all* experience stored? The holographic model says, yes, all experience is "stored" in the sense that we are inherently four-dimensional beings, that the holographic field is four-dimensional, that the flow of time is indivisible. In principle, given the right precision of modulation, i.e., the precise reconstructive wave, any event in the past should be capable of reconstruction. There is any number of forms of evidence of this preservation. Oliver Sacks (*The Man Who Mistook His Wife for a Hat*), for example, describes "Martin A.," suffering from a form of brain damage, who could nevertheless quote verbatim *Grove's Dictionary of Music and Musicians* – any of its six thousand pages. He heard these quotes in his father's voice – memories from the long hours his father devoted to reading and sharing the history of music with his beloved, though handicapped son. Sacks' retardate "Twins" could, given a date in their lives anywhere after roughly the age of four, give its details in total – the weather, the political events of which they might have heard, personally related events – as though they were simply reviewing a vast panorama unfolding before their inward eye. Kotre (*White Gloves*) reports the feats of some Jewish scholars, discovered in Poland around the turn of the century, who had memorized the entire contents of the Talmud, twelve volumes of thousands of pages. In demonstrating their ability, they would ask a volunteer to open the Talmud to any page. The volunteer would then take a pin and touch it to one of the words on the page, any word at all. The scholar would then ask the people in the room to call out other pages. Without looking he would tell them what words were in the same position as the pin on those other pages. The people could check out his accuracy by pushing the pin through the pages. Cases were documented in which the scholar never failed.

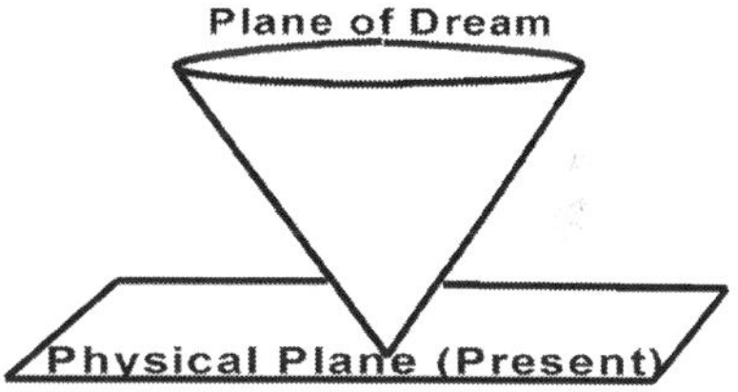

Figure 3.15. Bergson's Planes of Memory

Such phenomena are troublesome for the prevailing "abstractionist" trend in memory theory, but more experimentation is needed to prove the concept of complete recording of experience. Hitherto, however, I think it can be said that it has not been taken seriously. Nevertheless, the problem of obtaining the right precision of modulation to reconstruct any given event is a real one in this holographic theory. Further, it is doubtful that the concept of simply moving the brain into the right modulation pattern is a real basis to explain *all* these phenomena. Charles Tart (*Altered States of Consciousness*) reported for example that Aldous Huxley, author of *Brave New World* and 1984, could enter a meditative state in which he could view any page from any book he had ever read. Bergson made it quite clear that we are dealing with a four-dimensional being and aspects of consciousness that modern science has not yet addressed. In a case like Huxley's, the mind is abstracting from the present as far as possible, moving away from the perception-action state of the brain and deeply into the past.

Arthur Glenberg, a memory theorist, would later term this ability to ignore the call of the present perceptual array "suppression."[47] Bergson expressed this "movement of the mind" in terms of the diagram of Figure 3.15. The point of the cone is the most "focused" or concentrated point of the mind in the physical plane, completely concentrated on action. At the most spread-out plane of the cone, we have the realm of dream, of reverie, of the pure memory of experience. Between are various degrees of "focus," and indeed the computer scientist Gelernter used just this term to express the phenomenon as he observed it in *The Muse and the Machine*. Near the highly "focused" end or point, observed Gelernter, thought is abstract, conceptual. The origin of this operation (abstraction) of taking an entire "stack" of memories to examine one aspect of all them (an invariant) is one of high focus. This is in essence, the birth of concepts.

Even Larger Considerations

My wife, as I earlier related, has a previous California road driving experience of twenty years earlier return when driving along a certain curving section of Milwaukee roadway with its unique flow field. There is a memory retrieval question concerning retrievals of experience from even vaster distances of time. This particular question is of interest to those with a belief in or a curiosity about the remembrance of past lives from former reincarnations. A friend of mine who lives near Pittsburgh, Pennsylvania, while walking up a driveway which sweeps up a hill towards a large, Tudor style mansion, suddenly had an experience return of approaching a similar home, this time her own, from an apparent earlier life and time in medieval England. Is this possible?

When the dogma of neuroscience is that all experience is stored in the brain, a Dawkins or a Dennett can again look bemusedly at such reports, just as they would look askance at Douglas E. Harding. If all memory is stored in the brain, it is rather difficult to believe that memories of past lives get stuffed in there as well. Memory experiences of past lives can be safely written off as just more illusions generated by the brain. Unfortunately, it is this confident assurance that is an illusion.

I say again, the thesis that all experience is stored in the brain has never been more than this: an hypothesis. The history of the subject is dotted with absurd hypotheses, from the precise "maps" of the 1800's where memories were located in various areas in the brain, to the locations of the storage of "images" – visual and auditory . All these are now defunct. The questionable assertions continue today. There are theories of the storage of static features. As we have seen, there is no clue how these static features could be reassembled as dynamic events to reform experiences. There are theories that only certain events or certain aspects of events are stored, but absolutely no criteria for how such a selection is made.

Penetrating any possible principles of reincarnation is far beyond the science of today. However, the reconstructive wave model we have discussed here has no limits on what can be reconstructed from the past. The deeper question may be why there are such limits as there are, i.e., as Bergson asked, why is memory limited? But in any case, the theory is at least open to such possibilities, not forever closed on the basis of an unproven, but dogmatic hypothesis.

Chapter III: End Notes and References

1. Barsalou, L. W. (1993). Flexibility, structure and linguistic vagary in concepts: Manifestations of a compositional system of perceptual symbols. In A Collins, S. Gathercole, M. Conway, & P. Morris (Eds.), *Theories of Memory,* New Jersey: Erlbaum.
2. Mayes, A. R., & Roberts, N. (2001). Theories of episodic memory. In A. Baddeley, M. Conway, & J. Aggleton. *Episodic Memory*. New York: Oxford University Press, 86-109.
3. Robbins, S. E. (2006). On the possibility of direct memory. In V. W. Fallio (Ed.), *New Developments in Consciousness Research.*. New York: Nova Science Publishing.
4. In truth, it took courage for the times: Alan Paivio, *Imagery and Verbal Processes,* 1971.
5. Jenkins, J. J., Wald, J., & Pittenger, J. B. (1978). Apprehending pictorial events: An instance of psychological cohesion. *Minnesota Studies of the philosophy of science,* Vol. 9, 1978.
6. Bingham, G. P. (1993). Perceiving the size of trees: Form as information about scale. *Journal of Experimental Psychology: Human Perception and Performance*, 19, 1139-1161.
7. Gray, R. and Regan, D. 1999. Estimating time to collision with a rotating nonspherical object. In M. A. Grealy and J. A. Thomson (eds.), *Studies in Perception and Action V.* New Jersey: Erlbaum.
8. Pittenger, J. B., & Shaw, R. E. (1975). Aging faces as viscal elastic events: Implications for a theory of non rigid shape perception. J*ournal of Experimental Psychology: Human Perception and Performance* 1: 374-382.
9. The many experiments of Jennifer Freyd and associates can easily be seen to fit this parametric sensitivity.
   Freyd, J.J. (1987). Dynamic mental representations. *Psychological Review, 94*, 427-438.
   Freyd, J.J., & Finke, R. A. (1984). Representational momentum, *Journal of Experimental Psychology: Learning, Memory and Cognition. 10*, 126-132.
   Freyd, J. J., Kelly, M. H., & DeKay, M. L. (1990). Representational momentum in memory for pitch. *Journal of Experimental Psychology: Learning, Memory and Cognition, 16*, 1107-1117.
   Finke, R. A., & Freyd, J.J. (1985). Transformations of visual memory induced by implied motions of pattern elements. *Journal of Experimental Psychology: Learning, Memory and Cognition*, 11, 780-794.
10. Engelkamp, J. (1998). *Memory for Actions.* East Sussex: Psychology Press.
11. Zimmer, H. D., Helstrup, T., & Engelkamp, J. (2000). Pop-Out into Memory: A Retrieval Mechanism That is Enhanced with the Recall of Subject-Performed Tasks. *Journal of Experimental Psychology: Learning, Memory and Cognition*, **26**, 3, 658-670. (Quote on p. 669).
12. Engelkamp, (1998), op. cit.
13. Kim, N., Effken, J., Carello, C. (1998). Perceiving the severity of contact between two objects. *Journal of Ecological Psychology, 10*, 93-127.

14. Amazeen, E. (1997). The effects of volume on perceived heaviness by dynamic touch: With and without vision. *Journal of Ecological Psychology, 9*, 245-263.
15. Jenison, R. (1997). On acoustic information for motion. *Journal of Ecological Psychology, 9*, 131-151.
16. Steenbergen, B., van der Kamp, J., & Carson, R. G. (1997). Spoon handling in two-four year old children. *Journal of Ecological Psychology, 9*, 113-129.
17. Flynn, S.(1994). The perception of relative mass in physical collisions. *Journal of Ecological Psychology, 6*, 185-204.
18. Cabe, P., & Pittenger, J. (1992). Time-to-topple: Haptic angular tau. *Journal of Ecological Psychology*, 4, 241-246.
19. Kinsella-Shaw, J., Shaw, J., Turvey, M. T. (1992). Perceiving "walk-on-able" slopes. *Journal of Ecological Psychology, 4*, 223-239.
20. Pittenger, J. B. (1990). Detection of violations of the law of pendulum motion: Observer's sensitivity to the relation of period and length. *Journal of Ecological Psychology, 2*, 55-81.
21. Cabe, P. A. and Pittenger, J. B. (2000). Human sensitivity to acoustic information from vessel filling. *Journal of Experimental Psychology: Human Perception and Performance* 26: 313-324.
22. Vicente, K. J., & Wang, J. H. (1998). An ecological theory of expertise effects in memory recall. *Psychological Review*, 105 (1), 33-57.
23. Roediger, H. L., Balota, D., & Watson, J. (2001). Spreading Activation and Arousal of False Memories. In H. L. Roediger III, J. Nairne, I. Neath, A. Surprenant (Eds.), *The Nature of Remembering: Essays in Honor Robert G. Crowder*. Washington, D. C.: American Psychological Association.
24. Roediger, H. L, & McDermott, K. B. (1995). Creating false memories: Remembering words not presented in lists. *Journal of Experimental Psychology: Learning, Memory and Cognition, 21*, 803-814.
25. McKoon, G. and Ratcliff, R. (1992). Spreading activation versus compound cue accounts of priming: Mediated priming revisited. *Journal of Experimental Psychology: Learning, Memory, and Cognition,* 18, 1155-1172.
26. Lustig, C., & Hasher, L. (2001). Implicit memory is not immune to interference. *Psychological Bulletin, 127*, 618-628.
27. Cabe, P. A. and Pittenger, J. B. (2000), op. cit..
28. Robbins, 2006, op. cit.
29. Rogers, T., & McClelland, J. (2008). Précis of: Semantic Cognition: A Parallel Distributed Processing Approach. *Behavioral and Brain Sciences.*
30. Kugler, P. & Turvey, M. (1987). *Information, Natural Law, and the Self-assembly of Rhythmic Movement.* Hillsdale, NJ: Erlbaum.
31. Kingma, I., van de Langenberg, R., & Beek, P. (2004). Which mechanical invariants are associated with the perception of length and heaviness on a nonvisible handheld rod? Testing the inertia tensor hypothesis. *Journal of Experimental Psychology: Human Perception and Performance*, 30, 346-354.
32. Robbins, S. E. (2008). Semantic redintegration: Ecological Invariance. Commentary on Rogers, T. & McClelland, J. (2008). Précis on *Semantic Cognition*: *A Parallel Distributed Processing Approach. Behavioral and Brain Sciences*, 726-727.

33. Ingerman, P. (1966). *A syntax-oriented translator.* New York: Academic Press.
34. French, R. M. (1990). Sub-cognition and the limits of the Turing Test. *Mind, 99*, 53-65.
35. French, R. M. (1999). When coffee cups are like old elephants, or why representation modules don't make sense. In A. Riegler, M. Peshl, & A. von Stein (Eds.), *Understanding Representation in the Cognitive Sciences*, New York: Plenum.
36. French, 1999, p. 94.
37. Gentner, D. (1983). Structure-mapping: A theoretical framework for analogy. *Cognitive Science, 7*(2), 155-70.
38. Chalmers, D. J., French, R. M., & Hofstadter, D. (1992). High level perception, representation and analogy: A critique of artificial intelligence methodology. *Journal of Experimental and Theoretical Artificial Intelligence, 4*(3), 185-211.
39. Langley, P., Simon, H., Bradshaw, & G., Zytkow, J. (1987). *Scientific Discovery: Computational Explorations of the Creative Process.* Cambridge: MA: MIT Press.
40. Dietrich, E. (2000). Analogy and conceptual change, or you can't step into the same mind twice. In E. Dietrich & A. B. Markman (Eds.), *Cognitive Dynamics: Conceptual and Representational Change in Humans and Machines*, New Jersey: Erlbaum.
41. Doumas, L., Hummel, J., & Sandhofer, C. (2008). A theory of the Discovery and Predication of Relational Concepts. *Psychological Review*, 115, 1-43.
42. After Doumas, Hummel & Sandhofer, 2008.
43. Crowder, R. G. (1993). Systems and principles in memory theory: another critique of pure memory, in A Collins, S. Gathercole, M. Conway, P. Morris (Eds.), *Theories of Memory*, New Jersey: Erlbaum.
Goldinger, S. (1998). Echoes of echoes? An episodic theory of lexical access. *Psychological Review, 105*(2), 251-279.
44. Gelernter, D. (1994). *The Muse in the Machine: Computerizing the Poetry of Human Thought.* New York: Free Press.
45. Katz, J. J., & Fodor, J. A. (1963). The structure of semantic theory. *Language, 39*, 170-210.
46. Gayler, R.W. (2003). Vector Symbolic Architectures answer Jackendoff's challenges for cognitive neuroscience. In Peter Slezak (Ed.), ICCS/ASCS International Conference on Cognitive Science (pp. 133-138). Sydney, Australia: University of New South Wales.
47. Glenberg, A. M. (1997). What memory is for. *Behavioral and Brain Sciences*, 20:1-55.

CHAPTER IV

Why Robots Plead for Explicit Memory

They believed that they had understood and explained an intellectual act if they could break it down into its simple components – and they held it to be evident or dogmatically certain that these components could consist in nothing other than simple sense impressions...But it was precisely this fundamental vision which inevitably removed them in theory from the peculiar principle and problem of the symbolic.

- Ernst Cassirer, *The Philosophy of Symbolic Forms*[1]

It is the lasting achievement of the Bergsonian metaphysic that it reversed the ontological relation assumed between being and time.

– Ernst Cassirer, *The Philosophy of Symbolic Forms*[2]

Robotics and Consciousness

In 1997, IBM's chess playing machine, Deep Blue, defeated the human world champion, Gary Kasparov, 3½ games to 2½ games. In 2006, another chess computer, Deep Fritz, defeated world champion Vladimir Kramnik, four games to two. The computers have reached this achievement due to increasing processing power, sophisticated algorithms or methods to calculate the long range results of a given move on the board on the order of a dozen moves in the future, and the addition to the computers' memories of a large catalog of board configurations, on the order of 50,000, and the normal game-results that tend to ensue once such a pattern occurs. A current board configuration in a game can be matched, in this memory, against this large catalogue of board configurations and the results that tended to occur given such a pattern on the board. The end-result of all these improvements, together with the addition of yet more processing power, is making the machine "players" – which are unfazed by fatigue – nearly invincible.

The image of machine dominance at chess is taken as a harbinger of what is in store for humanity. Things appear ominous. It lends credence to Ray Kurzweil's prediction of a "singularity" (*The Singularity is Near*). Compared to the turtle-like, 100 meters per second transmission speeds of neurons in the human brain, the circuits of robots will support near-light speed transmission velocities. Every good architectural aspect of the human brain, Kurzweil asserts, will be exploited and amplified, doubled, tripled in power by electronic components. We will end incorporating these components into our own body, amplifying our intelligence. But the machines, unfettered by biological limitations, will assuredly continue to evolve more quickly. The year of this tipping point – the singularity – Kurzweil boldly predicts, will be 2045.

Chess, however, would be the wrong subject from which to take encouragement. It is a "micro-world." It is a very constrained "event." The game has a very finite set of pieces, and a finite set of moves each piece can make. It is the permutations and combinations from this finite set – the number of possible games that can play out – that is huge. But the elements and "motions," that is, simply the change of positions, in this constrained micro-world can be represented as internal symbols in a computer. It is perfect grist for the abstract space underlying computing. The complexity here arises from the infinite variety of patterns these pieces and motions can take in their constrained space – all with a final goal of a "checkmate" configuration. .

The creation of a micro-world was an early strategy of AI to conquer the problem of commonsense knowledge of the world. In the early 1970's, Terry

Winograd, then a graduate student at MIT, constructed a program called SHRDLU as an approach to computer understanding of natural language sentences.[3] The sentences input to the program were restricted to a very small, well-delineated, simple domain: a micro-world. This world, for SHRDLU, consisted of nothing but a set of blocks on a tabletop displayed on a computer screen, together with a hand for moving them about. The blocks came in different shapes and colors, and could be stacked upon one another, e.g., a pyramid atop a cube. The computer kept a continuously updated database of which blocks were where, whether a pyramid was atop the cube, the cylinder next to such and such a block, which block was held in the hand, etc. It could answer questions such as:

User:	What color is the block on the red brick?
SHRDLU:	GREEN.
User:	Shape?
SHRDLU:	PYRAMID.
User:	Grasp the pyramid.
SHRDLU:	I DON'T UNDERSTAND WHICH PYRAMID YOU MEAN.

(There are three pyramids in the scene.)

The micro-world, however, is again a predefined "object language" specifying the problem environment, and the relevant features thereof, in which SHRDLU works. The feature set of this "world" is fixed – the properties of the blocks, their move-ability, their ability to stand on another or not (the cube can't sit on top of the pyramid), requirements for support, etc. The rest is again simply a form of "proof" procedure taking place within this fixed symbolic framework.

SHRDLU, given its complexity and programming ingenuity, was rightly hailed as a great achievement, but wrongly seen as a precursor of great the things to come. Winograd himself, intimate with its limitations, repudiated the entire approach in relatively short order. With Thomas Flores, he moved to the concept of the "situatedness" of a being in the world, and rejected the view of cognition as symbolic manipulation of internal representations that are understood as referring to objects and properties in the "external" world. Following Martin Heidegger's philosophy of being-in-the-world (*Being and Time*, 1927), Winograd and Flores noted:

> Heidegger makes a more radical critique, questioning the distinction between a conscious, reflective, knowing

> "subject" and a separable "object." He sees representations as a *derivative* phenomenon, which occurs only when there is a breaking down of concernful action. Knowledge lies in the being that situates us in the world, not in reflective representation.[4]

There is only one way to understand this rather abstruse passage. Winograd's "situatedness" is precisely Bergson's union of subject and object "in terms of time." Heidegger was well aware of Bergson. In Bergson's model of perception, there are no internal "symbols." There are no internal symbols for the "blocks" of the external micro-world that SHRDLU is "looking" at; there are no internal symbols for the "buzzing" fly or the rotating cube. The "symbols" *are the objects*, precisely where they are *specified* to be, externally, in the flowing matter-field. The "symbols" *are* the blocks, right where they appear to be; the "symbol" *is* the fly right where he is buzzing; the symbol *is* the cube, right where it is rotating. What we normally term "symbols" – the symbols or images we use in representing the world – only arise when we have to go back into memory, when we have to redintegrate, when we have to recall an event or events, when we have to "represent" a piece of the world to ourselves via an image of memory. This "representing" state is now Heidegger's "breakdown of concernful action," for we are now reflecting, not *acting* without representation on the concrete objects of the world, as when we are simply and naturally picking up the blocks, reaching for the fly or driving down the flowing road.

This is why robotics is inescapably tied to the problem of consciousness, and in turn, to the relation of subject and object, and thus to the flow of mind in a non-differentiable time. Current robotics theorists, however, already wish to claim their devices are conscious, or nearly so, and can be made so with "new" architectures that are unrecognizably different from machines that already exist. A 2007 issue of the *Journal of Consciousness Studies* featured a number of such theorists on the subject of machine consciousness. The level of non-awareness on the nature of the true problems their robots face on their way to becoming "conscious" is vast.

The Robots Cannot "See" the Problems

One theorist, Julian Kiverstein, begins with what, on the surface, seems an awareness of the "coding problem" facing any self-respecting robot. He notes:

> Indeed when we imagine a zombie, aren't we imagining a creature whose existence is much like that of a robot?

> We imagine something that makes all the right moves and produces all the right noises, *even though all is dark within*: the machine has no inner mental life.[5]

This is a compelling intuition of the problem – this darkness within. But Kiverstein takes inspiration from studies using a Tactile Vision Substitution System (TVSS). With a TVSS device, optical images are picked up via a head-mounted camera and the signals are transduced to an array of stimulators on a person's skin. Using a TVSS, people who have become blind report something very akin to visual experiences of external objects. Critical to this is that the person actually controls the camera (which is to say it is actually mounted on the person's head). With this setup, he or she receives direct feedback as to how the optical information changes as the person's movements change, which is only to say that it is this connection to movement and change that is enabling the isolation of the invariants from the variants, i.e., the detection of the invariance laws. From such studies, Kiverstein concludes that all that is required to account for conscious perception is, "that the machine understand the patterns of dependency that hold between sensory stimulation and movement."

Magically here, the coding problem has again been solved. The optical stimulation, now transduced to tactile information playing on the skin, is yet a code for the external world. The action which the information is related to is only another form of code; it is not the external world. As per usual, there is absolutely nothing in this theory to explain how an *image* of the external world arises from either of these two codes or their combination. That such a TVSS setup is again allowing the brain to create the modulated wave patterns passing through the external holographic field and specifying a reduced form of image of the field, or in reality, a memory image – this we can understand. But Kiverstein is not saying this, and to make sense of this requires the framework on subject, object and time which we have seen in Chapters I and II. Simply assumed here also in Kiverstein's solution is the "primary memory" that underlies all perception – the rotating cubes, the buzzing flies – that allows this specification of a past motion of the external matter field. But the complete lack of awareness that we must account for the time-continuity of perception and therefore the memory that underlies it is endemic in these theories of robots and their possible consciousness.

To add to this confusion, Kiverstein's entrancement with the pure theory of "computation" or symbol manipulation underlying the computer metaphor allows him to state that "there is nothing special about our biological wetware." As long as we create a system which connects sensory information to action, we will have consciousness, whatever the materials, be it silicon, wire, fiber optic tubing, etc. Again, we have the misguided idea that employing logic and symbol manipulation is all that is required. Remember, Arnold the Terminator connects sensory information to action. He is still blind as a bat. Lost here is all notion that, indeed, like the AC motor, the actual dynamics, the forces and masses and fields of physics, e.g., the concrete form of a concrete wave passing through a concrete holographic field, may be all important.

But Kiverstein is not quite satisfied. This is not the ultimate solution after all. Something more is needed. In addition, there must be a form of "comparator" mechanism. The comparator constantly compares the result of an action by the person to a prediction of the results of the action. So there is always a prediction, then an action, then a comparison of the actual event to the results (Figure 4.1). For example, argues Kiverstein, an object will look smaller when seen from a distance, or larger when one moves closer, yet the size of the object is experienced as constant. "Thus one experiences the true size of an object across changing experiences because one has correct expectations about how an object will look at various distances." So, invoked here, likely unknowingly, is a 1700's theory of Bishop Berkeley on the nature of size constancy which Gibson transcended sixty years ago (noted in Chapter I). Being ignored is that size constancy is specified by the constant relation of the texture units occluded on a gradient by the height of the object as it moves back and forth – an invariant over time and over the flow field that is useful for specifying possible action. But Kiverstein, as most theorists of so-called robotic consciousness, is unaware of this. He is conceiving of the brain at a taker of "snapshots" of a flow field. This is to say, to him, time is a series of "instants." If, in each snapshot, there is a "match" between expected result and actual results, the match results in a form of self-awareness, i.e., consciousness. So it is, apparently, the comparator that supposedly explains self-awareness. But the comparator, as it proceeds instant by instant in this model of time, can have no continuity, no memory. Each "match" is simply a comparison of two symbols in another instant. Hence it implies no consciousness, no awareness whatsoever. Arnold does the comparator thing too.

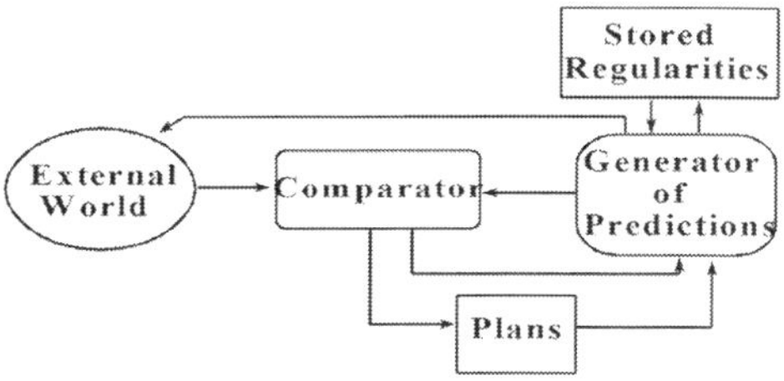

Figure 4.1. A "Comparator" System. Here images or snapshots, stored in memory, are matched against "current" images of the external world.

Yes, the "comparator" is simply a notion based on "samples" or snapshots of events, events such as our rotating cube. A snapshot is taken of the rotating cube and compared to a predicted snapshot. Imagine, sitting in your chair, watching the wobbly cube. Somehow, the robot is supposed to be generating predictions, sample by sample, of a flowing event. What is the time-scale of these samples? Each nanosecond? Each microsecond? We have seen in Chapter II, when discussing the wobbly cube, that there can be no such snapshots. The information for the form of the cube, whether wobbly or rigid, is not found in a single snapshot. There are no edges, no vertices, no flat surfaces in a snapshot. The fact is, this comparator conception is built upon a lack of awareness of the modern concept of velocity field flows underlying the form of the cube. And this failure of awareness of the theory and problem of perception is absolutely typical in the robotic literature.

Finally, not satisfied with this, Kiverstein introduces yet another ultimate ingredient. This is *agency*, or the intention to perform an act, or to plan. But invoking "agency" is again simply to beg the question, for "agency" too already implies consciousness. It implies at least an image and a thought of a goal that can be sustained over time – an *extended* time. The robotics theorist tends then to look for a more mechanistic method of supporting planning in a robot. For this, Kiverstein invokes "means-end" analysis. Where have we seen this notion before? Unfortunately, in 1972! Newell and Simon's General Problem Solver featured, front and center, "means-ends" analysis. The monkey used it to get to the bananas; GPS used it to solve logic problems. No one will accuse GPS or a computer running it of being conscious.

This is enough delving into the literature describing what it takes to create conscious robots. Kiverstein, whom I hope will pardon me for this focus upon his theory, is simply representative. In general, the literature is sorely unconscious of the problems it must solve. It ignores the problem of time and the true nature of the problem of perception, is highly leveraged on the illusory success of connectionism, is woefully neglectful of the science of ecological psychology, and as we shall see now, ignores the great problem of explicit memory. I will set out the criteria that must actually be met for a conscious "device" in Chapter VI.

Explicit Memory vs. Robotics

Explicit memory requires consciousness. What is termed *implicit* memory does not require consciousness. Though the distinction is ubiquitous in the memory literature, there is no understanding in the current theories why this is so, particularly why explicit memory requires consciousness or what this means. Remember, all current cognitive theories have no role for consciousness. Implicit memory as a concept arose with the discovery that amnesics can exhibit a form of learning – a learning of which they themselves are not explicitly aware. The original discovery of this was with the case of a famous amnesic, H.M.

In 1956, William Scoville removed the medial portions of both H.M.'s temporal lobes in an attempt to control an epileptic condition. After this operation, the nature of his conscious experience changed markedly. He displayed a severe Amnestic Syndrome, appearing unable to retain any further long-term memories. H.M. cannot remember events from one day to the next, in fact for far briefer intervals. He could retain a string of three digits, e.g., 824, by means of an elaborate mnemonic/rehearsal scheme for perhaps fifteen minutes. Yet five minutes after he had stopped and explained the scheme to experimenters, the number was gone. Remarkably, his same examiners can come into his room day after day, yet from day to day he does not remember ever having seen them. It was Milner who discovered that H.M. could learn perceptual or motor skills such as mirror-tracing. But, though steadily improving from session to session, he insisted, upon entering each practice session, seeing the mirror-tracing apparatus, etc., that he had never done this before. For H.M., by his own description, each day is "a day unto itself," without history.

Clive Wearing lives perhaps in an even shorter temporal frame. I first saw Clive discussed in Gray's (1995) paper modeling the hippocampus after a "comparator" of present vs. past event.[6] Clive is another individual, similar to H.M., with extensive bilateral damage to the hippocampus, amygdalae, etc. Clive, as H.M., is unable to remember previous events. He feels constantly that he "has just woken up," and keeps a diary in which this feeling is repeatedly recorded at periods of hours, even minutes. Statements are made such as, "Suddenly I can see in color," "I've been blind, deaf, and dumb for so long," "Today is the first time I've actually been conscious of anything at all." To Wearing, the environment is in a constant state of flux. A candy bar, even though held in his hand, but covered and uncovered by the experimenter, appeared to be constantly new.

While Clive Wearing and H.M. are at the extreme end of the Amnestic Syndrome, it is in such cases that the problem of the explicit is shown in stark relief. Clive and H.M. are the epitome of a consciousness whose scope is limited to the present. The fundamental dynamics are operative that support the specification of the "world-out-there," the multi-modal world of present experience – conscious perception. As in H. M. and "824," there is even extremely short-term memory. But as soon as we move to the slightly more remote past, beyond the time-scale of perception and events held by rehearsal, a whole set of higher harmonics as it were, beyond the fundamental, are no longer part of the "chord" supported by the dynamically transforming brain.

We have seen the mechanics of redintegration. Now we see that this mechanics must include some of these overtones, beyond the fundamental, to explain direct recall of events as explicitly *past* events, i.e., to *localize* these events as events in one's past. H.M. approaches for the *nth* time the mirror tracing apparatus. The same flow field is specified as he moves to the device, moving across the same texture gradient specifying the same floor, the same table surface, the same optico-structural transformation of the device as he moves towards it, the same action being carried out. Everything required for redintegration appears present – the same event pattern, the same perceptual information. The redintegrative dynamics should be perfectly supported. The phenomenon of priming, which amnesics still demonstrate to have, indicates that it is. But there is no *explicit* memory of ever having done, seen or experienced this mirror-tracing thing before. H.M. can redintegrate 'til the cows come home, he won't remember an event as an event in his past. For a theory of consciousness, the problem is startling, with strange symmetries to the problem of perception. We have all the neural dynamics supporting perception and redintegration operating, yet just as in the missing "qualia," there is here a mysterious ingredient missing – the ingredient that makes the past event consciously remembered.

The prominent memory theorist, Lawrence Weiskrantz, argued that in the case of the amnesic, a wide range of skills is capable of being learned and retained, not just the procedural (as in mirror tracing), in fact, nearly anything.[7] The list includes:

- New rule learning for verbal paired associates.
- A new artificial grammar.
- Classification of novel drawings relative to drawings previously seen.
- Mathematical problems.
- Even the answers to anomalous sentences, e.g., "The haystack was important because the cloth ripped." Answer: parachute.
- Word processing (computer) skill and its associated vocabulary.

There appears to be no limit, no limit, that is, as long as the task does not require the amnesic to place an event in the past. Weiskrantz expressed these amnesic-capable tasks as tasks which do *not* depend on a "joint product" relating present to past (current event x stored event). The amnesic simply cannot perform this product, this fundamental comparison – *present x past.*

The Problem of the Explicit

The memory ability that a Wearing, H.M., or amnesic in general does indeed preserve is generally subsumed under the concept of the *implicit*. Since the first coining of this now ubiquitous explicit/implicit distinction, I think it safe to say that there has been relatively little theory on the true nature of the difference, i.e., on the nature of the *past x present* product of Weiskrantz. In truth, to my knowledge, the only journal article attempting this is my own.[8] In general, the explicit is associated with consciousness, the implicit is not. Little is discussed beyond this, with emphasis generally on conditions or variables inducing or affecting implicit memory performance.

In general, it is the view that loss of explicit memory is the result of damage to an explicit memory "system," generally considered to involve the pre-frontal, cortical areas and medial-temporal lobes, but just why its proper functioning yields a *conscious* remembrance is vague. The critical difficulty, for consciousness, is why – even in a fully functional "explicit memory system" in which all retrieval mechanisms are fully operative, be they connectionist neural nets with their firing patterns, or the operations of cognitive symbolic programs – should these operations result in a *consciously* experienced remembrance as opposed to just another "implicit" operation. This is the question: Just what is required for a past event to be consciously experienced as an event localized in one's past? What must such a system actually involve? This is what we will now address.

The COST of the Explicit

It is generally accepted that it requires roughly two years for the child to develop explicit memory.[9] In fact, this period coincides with something termed *childhood amnesia*, namely, the inability to remember the early events of childhood. Something is occurring in the brain, as a dynamic self-organizing system, over these two years which eventually allows the brain to achieve a complex dynamic state. This does not mean, by the way, that experience is not "stored" during these two years. It means that the brain, far more physically complex than anything of which the robotics theorist has ever conceived, requires two years of dynamic organizing to achieve this ability to perform a *past x present* product. It apparently has not occurred to the robotics theorist to ask what is going on during this time and why the brain must go about it in this way, whereas their brilliant robots find it unnecessary, performing it with a simple if-check comparison of symbols.

To the great child developmental theorist, Piaget, this dynamic development in the brain's organization is underpinning the birth of a correlated set of fundamental concepts – Causality, Object, Space and Time. I have termed this the "COST" of explicit memory, for simultaneously this complex is giving birth to, and required for, the explicit memory of events – in Piaget's terms, the ability to "localize events in time." This too underpins the development of the ability to symbolize.[10]

I am not going to trace in any detail Piaget's description of the development of this ability through his six "stages."[11] These stages, by the way, are now seen as natural punctuations in the development of a self-organizing dynamic system.[12] To summarize, the concepts of COST form together an interrelated, supporting group, and together they grow from the inter-coordination of the child's actions. As Piaget describes it from the perspective of the development of one of these, namely from the concept of the *object*, from a mere *extension* of the child's activity, the object is gradually dissociated from his activity. Resistance initiates this dissociation, e.g., obstacles or complications in the field of action as in the appearance of a screen obscuring a favorite toy.[13] Action gradually becomes a factor among other factors, and the child comes to treat his own movements on a par with those of other bodies. This is coordinate with the development of causality, for the child has now moved from a state where everything that happens to him is simply part of his own subjective being, to where he, as an object, sees himself as equally subject to forces from other, external objects. This in turn is coordinate with the child's beginning ability to order experienced events in an objective series in an objective time.

Piaget gave a classic example of an event (in the Sixth Stage) where his 19 month-old daughter, Jacqueline, first demonstrated an achievement of a symbolic representation, a representation reliant on explicit memory of the past or a "localization of events in time."

> Jacqueline (19 months) picks up a blade of grass which she puts in a pail as if it were one of the grasshoppers a little cousin brought her a few days before. She says "Totelle

> [sauterelle, or grasshopper] totelle, jump, boy [her cousin]." In other words, perception of an object which *reminds her symbolically* of a grasshopper enables her to evoke past events and reconstruct them in sequence (emphasis added).[14]

The present event (the piece of grass) now functioned within what I term an *articulated simultaneity*, for it served both as itself and as a symbol of a past event, assuming both one meaning then another, yet within a global, temporal whole or state of consciousness, all underpinned by the non-differentiable flow of time in which the brain dwells. The amnesic, however, has lost or suffered damage to the brain's ability to support this complex dynamic state. Explicit memory is bound to the development of the symbolic – the ability to employ a form (the piece of grass) abstracted from our experience to represent another aspect of experience in the past (the grasshopper) – what Cassirer would term the "symbolic function."[15] The symbolic function carries this aspect of an articulated simultaneity – this "oscillation" between two elements within a temporal whole. This tells us that the symbolic itself is a problem in the relation of mind to time.

The Simultaneity of the Symbolic Mind

I wish to underline the critical aspect of this symbolic representation, namely the articulated simultaneity just noted above. It is not only the events of the past that become symbolic. For the philosopher, Ernst Cassirer, it is the events of perception as well. The pathologies, he thought, indicate that, "the contents of certain sensory spheres seem somehow to lose their power of functioning as pure means of representation…"[16] Some aphasics cannot make a simple sketch of their room, marking in it the positions of objects. Many patients can orient themselves on a sketch if the basic schema is already laid down, e.g., the doctor prepares the sketch and indicates by an X where the patient is. But the truly difficult operation is the spontaneous choice of a plane as well as a center of coordinates. Thus, one of Head's patients would express his problem as the "starting point, but once it was given him everything was much easier."[17]

The same principle operates in the aphasic's dealing with number and time. One patient could recite the days of the week or the months of the year, but given an arbitrary day or month, could not state what came before or after. Though he could recite the numbers in order, he could not count a quantity. Given a set of things to count, he could not progress in order, but frequently went back again. If he had arrived, for example, at "six," he had no comprehension that he had a designation for the quantity thus far achieved, i.e., a cardinal number. When asked which of two numbers are larger, say 12 or 25, many aphasics can do so only by counting through the whole series, determining that in this process the word for 25 came after the word for 12.

As Cassirer notes, "where quantity no longer stands before us as a sharply articulated multiplicity, it cannot be strictly apprehended as a unity, as a whole built up of parts."[18] But to achieve this, every number must carry a *dual* role. Thus to find the sum of 7 + 5, or the difference 7 - 5, the decisive factor is that the number 7, while retaining its position in the first series of 7, now is taken in a new meaning, becoming the starting point of a new series where it assumes the role of zero (Figure 4.2). Again, as in the representation of space, we have a free positing of a center of coordinates. "*The fundamental unities must be kept fixated, but precisely in this fixation be kept mobile, so that it remains possible to change from one to the other.*"[19] The number 7 must maintain its meaning as 7, yet *simultaneously* assume the meaning of zero. This is a pure problem of representation in time. More precisely, it is a problem of representation in an *extended* time supporting true simultaneity.

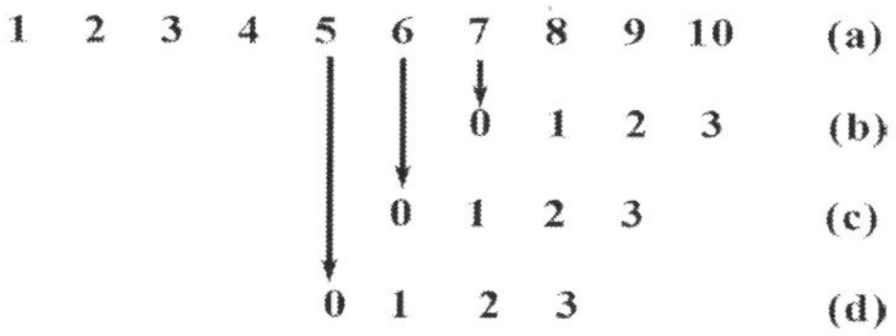

Figure 4.2. The problems: 7 + 3 = 10, 6 + 3 = 9, 5 + 3 = 8. The number 7 functions simultaneously as zero in (b), 6 functions as zero in (c), and 5 as zero in (d). (After Cassirer, 1929/1957)

This extends to the sphere of action, in the apraxias. One patient (Gelb and Goldstein, 1918) could knock on the door if the door was within reach, but the movement, though begun, was halted at once if he was asked to move one step away from the door. He could hammer a nail if, hammer in hand, he stood near the wall, but if the nail was taken away and he was asked to merely indicate the act, he was frozen. He could blow away a scrap of paper on the table, but if there was no paper, he could not blow. This is not the loss of memory images, as was once widely held. He just blew the paper, how could the image have been lost? It is not a failure to create a sensuous "optical space" as Gelb and Goldstein thought. He is staring at the door, and still cannot act. It is the inability to create an abstract, free space for these movements.

> For this latter is the product of the "productive imagination": it demands an ability *to interchange present and non-present*, the real and the possible. A normal individual can perform the movement of hammering a nail just as well into a

> merely imagined wall, because in free activity he can vary the elements sensuously given; by thought he can exchange the here and now with something else that is not present… [this] requires a schematic space.[20]

This extends to analogy. Thus the psychic blindness patient of Gelb and Goldstein was utterly unable to comprehend linguistic analogy or metaphor. He made no use of either in his speech, in fact, rejected them entirely. He dealt only in realities. Asked by Cassirer, on a bright, sunny day, to repeat, "It is a bad, rainy weather day," he was unable to do so. Again, analogy requires precisely the same achievement of representation. To say, "The spoon is a catapult," (after using it to launch a pea) the spoon must be taken simultaneously in two different modes. One must place oneself simultaneously now in one, now in another meaning, yet maintain a vision of the whole. It is this "articulation of one and the same element of experience with different, equally possible relations, and *simultaneous* orientation in and by these relations, [that] is a basic operation essential to thinking in analogies as well as intelligent operation with numbers…"[21] While we noted that computer models that have been advanced for this analogic capacity of mind simply give away the problem with their predefined "features," this simultaneous articulation represents yet another dimension that these models have not even a thought of supporting.

It is this simultaneous articulation that supports little Jacqueline and her "grasshopper." Again, as Cassirer argued, there is an interchange of present and non-present, i.e., of present and past. The little piece of grass can now be a symbol; it can be simultaneously both a piece of grass – a *perceptual image* – and a "grasshopper" which once she saw hopping around (something that is also already – an analogy). When H.M. looks at the mirror tracing apparatus, should it not be simultaneously both the present apparatus and the past apparatus upon which he once worked? Sitting on my porch, the wind chimes hanging near by instantly become a "symbol," simultaneously both of themselves and the memory of buying other chimes for my wife for an anniversary. All such surrounding objects can become *symbolic* of the past. But now we see this apparently simple ability through the underlying complexity of the dynamic that must support it.

The physical organism spends several years and a great deal of effort following this "law of evolution" or developmental trajectory to produce this dynamical possibility. In the sphere of explicit memory, and in the spheres Cassirer discussed – sketching, analogy, numbers, time, voluntary actions – an essential feature of the underlying dynamic in each is the articulated simultaneity supporting the symbolic nature of these functions. This can apparently be disrupted in each. In the case of amnesia, I would argue, damage to the neural circuitry underlying the concrete dynamics required for explicit recall is effectively disrupting the ability to sustain the articulated simultaneity in time necessary for the present event to be simultaneously in a symbolic relation to the past.[22]

Remembering Sticks and Flows

Cassirer's "articulation of one and the same element of experience with different, equally possible relations" is found in numerous examples described by Piaget on the course of child development. Again it underlies representation. Let me take at a more advanced point on the developmental trajectory, a point which also illustrates what can be termed the *dynamical lens on events* inherent in Piaget's view. Piaget describes a simple memory experiment with children aged 3 to 8.[23] They are shown a configuration of ten small sticks (Figure 4.3 [A]). They are asked to have a good look so they can draw it again later. A week later, without having seen the configuration again, they are asked to draw what they were shown before. Six months later they are asked to do the same thing.

In the one week interval case, the reconstruction of the event is dependent, in Piaget's terms, on the "operational schemata" to which the child assimilated the event. The memory is dependent, in other words, on the child's current ability to coordinate actions. At around 3-4, the series is reproduced as in Figure 4.3[B]. Slightly older children (4-5) remember the form in Figure 4.3[C]. Figure 4.3[D] is a slightly more advanced reproduction, while at 6-7, the child remembers the original series. After six months, as Piaget describes, children of each age group claimed they remembered very well what they had seen, yet the drawing was changing. The changes generally, with rare exceptions, moved in small jumps, e.g., from A to C, or from C to D, or from D to A.

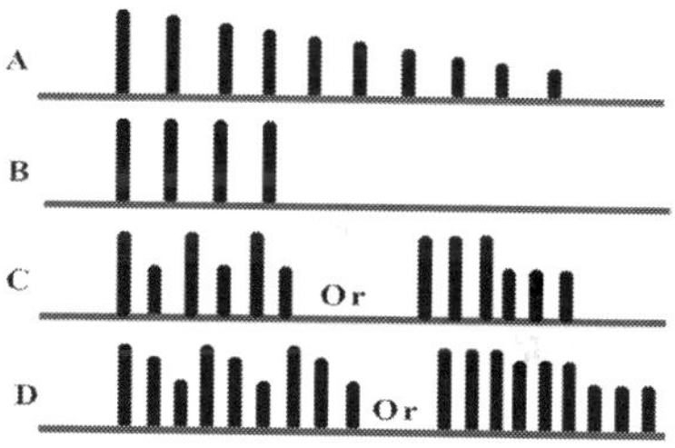

Figure 4.3. Stick Series

The drawing of a series – the process of seriation – requires "concrete" operations as Piaget terms them. To order a simple series, A, B, C (where A is longer than B, B longer than C), one must simultaneously relate or coordinate the height of B relative to A, the height of B relative to C. This is to say, B must be represented as simultaneously both smaller than A and longer than C – simultaneously both one relation and another. This is fundamentally based on the inter-coordination of action. As the child follows this developmental trajectory, Piaget argues, the available dynamics supports successively different "de-codings" of the memory, i.e., the events are reconstructed with an increasingly sophisticated logical structure. The past event is seen through this dynamical lens. But simultaneously then, when the series is a perceived event, it is being perceived through this same dynamical lens. The events gain increasingly complex *symbolic* structure.

This is but one of a large number of such examples of events treated by Piaget where we see this developing dynamical lens, and the need for a *simultaneous* relation of magnitudes. In another, more dynamic event, the experimental apparatus consists simply of two differently shaped flasks, one placed atop the other, with a tap or valve between the two (Figure 4.4).[24] The top flask is initially filled with liquid. The child (age 5-9) is provided with a set of drawings of the two flasks (with no liquid levels filled in). At regular intervals or stages a fixed quantity of liquid is allowed to run from the top flask into bottom until the top is empty. Because the flasks are differently shaped, the one empties at a different velocity than that at which the other fills. At each stage, starting at the beginning, the child is asked, using a fresh drawing, to draw a line on each flask indicating the level of the liquid.

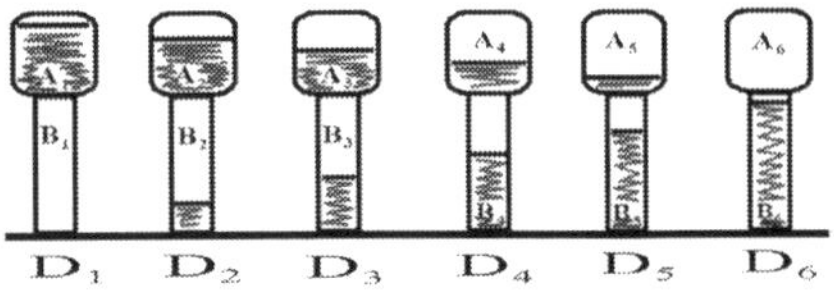

Figure 4.4. Two flows at different velocities. In a series of stages, the water is gradually emptied from the top beaker (A), while the lower beaker (B) gradually fills.

The drawings are now shuffled and questions ensue:

1) The child is asked to reconstruct the series, putting the first drawing made down, the next to the right, and so on.
2) Every sheet is now cut in half and the drawings shuffled. The child is again asked to put them in order.

The children again go through stages. At first they cannot order the uncut drawings (the D's of Figure 4.4). Then they achieve this (uncut) ordering but fail ordering the cut drawings. The child may come up with an arrangement of the cut drawings such as A_3, A_1, A_2, A_5, A_6 above B_1, B_5, B_6, B_3, B_2. Even with coaching, he cannot achieve a correct order. Beneath these failures lies a certain mental rigidity. The child is required here to move mentally against the irreversible, experiential flow of time. Regardless of the irreversible flow of the water, they must perform a reversible operation. They must construct a series A -> B -> C... where the arrows carry a *dual or simultaneous meaning* standing for "precedes" as well as "follows." *It is a pure problem of representation.* The child must also grasp the causal connections within and between the two flows. Ultimately, the children will surely and securely order this double series with its causal link as the operating principle. The operations involved in this coordination of two motions, participating in the logic of objects we call causality, underlie our "schema" of time. This schema, supporting our memory operations, our retrieval of events, is again a lens upon our experience. It is a cognitive capability that develops over several years, based on the coordination of actions, first sensori-motor, then mental – and something that must be subject to disruption via a traumatic injury or disease.

Failure to Net Piaget

The connectionist modelers have of course attempted to weigh in on Piaget's tasks. Rogers and McClelland, in support of their connectionist vision of the development of cognition, note that several connectionist models have been developed which can perform developmental learning tasks.[25] One such is the balance scale problem of Piaget. Here we have a small apparatus, like a small teeter totter, upon which the child can hang weights on each side to balance the "teeter totter" arm. The children gradually (again, in stages, over years) learn the principle of torque, i.e., the rule for balancing two differing weights (say a five gram weight and a ten gram weight) on each side, allowing compensation by varying the distances from the center of the scale. For example a five gram weight on the right, and two feet from the center, will balance a ten gram weight on the left, only one foot from the center (length x weight = length x weight, or 5 lbs x 2 ft = 10 lbs x 1 ft). Yet Quinlan et al., in a detailed analysis, have shown that in balance scale learning simulations, connectionist networks *never* in fact learn the principle of torque, nor do their internal weight representations ever approximate intermediate rules that correspond to the human phases of learning.[26] Here the mere *approximation* of connectionism to human developmental learning is unmasked.

There is a significance to this. Again it is that connectionist models are insensitive to the transformations and invariants defining events. Balancing a scale is an event. The full semantics of the principle of torque exists across a set of such events with varying weights and distances; there is again the perception of an invariance law (length x weight = length x weight) over these balancing transformations. It would be a quite feasible exercise to show that this will be true for nearly all Piagetian tasks. We shall see this forcefully in the tunnel-bead experiment of Piaget in Chapter VI, where again, as dynamic, self-organizing systems, the children require several years to master the task.

After seeing the requirements for mastering the coordination of flows discussed above, with the intrinsic reliance on the explicit/symbolic (or COST), the reader is now in position to appreciate the actual magnitude of the task that would truly face the connectionist networks in trying to simulate this development. These Piagetian tasks require explicit memory and the symbolic function. The connectionist network is approaching these tasks in a pure syntactic mode as a device (network) whose operations function logically in an abstract time. Without the benefit of a dynamics that is integrally part of the non-differentiable flow of time, such a network has no hope of modeling these tasks.

Failure to Mass-Spring Piaget

In Piaget's Stage Four (8-11 months), there is an interesting point in the growth of the explicit. Here the child begins to apply the fundamental "before-after" relation in time to *objects*. An object disappears behind a screen, but while perceiving the screen, the child retains the image of the object and acts accordingly. The child is now recalling events, *not merely his own actions*. But this is extremely unstable. The object, say a toy, is hidden originally at A, and found by the child at A (e.g., under a pillow). The toy is then moved in full view to B and hidden under another pillow (B). The child goes to B, but if the toy is not found immediately, goes back to A, where the *action* was originally successful. The child's memory is now such that he can reconstruct short (but only short) sequences of events independent of the self.

The searching phenomenon described above is commonly termed the *A-not-B* error in the developmental literature, and has been the subject of extensive research, and yes, even connectionist models.[27] It has been part of the continuing attempt to reduce Piaget's children to a pure mechanism, without the need of consciousness. Thelen et al., on the same agenda, follow a dynamic systems approach.[28] Here the attempt is to treat this by modeling the child's decision as a dynamic field which evolves continuously under the influence of the A-B task environment and the specific cue to reach to A or B (which must be remembered). This includes, after the child's first reach, the memory dynamics which "bias" the field on a subsequent reach. This is roughly equivalent to biasing a coiled mass-spring (a spring with a weight on the end) to release or spring out at a certain point. The field represents the relative activation states of the parameters appropriate to planning and executing a reach (that is, to releasing the mass-spring) in a specific direction to the right (A) or left (B).

But consider the reflections of the authors in trying to account for the reasons for change of, and the origins of, the value of the critical parameter in their model (*h*, or cooperativity) that determines a reach to A or to B:

> They learn to shape their hands in anticipation of objects to be reached and then to differentiate the fingers to pick up small items...They start to incorporate manual actions with locomotion such as crawling and walking. And they begin to have highly differentiated manual activities with objects of different properties, such as squeezing soft toys and banging noisy ones. It is a time of *active exploration of the properties of objects by acting on them* and of active exploration of space by moving through it.[29]

If one studies Piaget, one sees that this is exactly what he describes his children, Jacqueline, Lucienne and Laurent as doing – in detail, stage by stage. The above description simply strips out the embedding in the larger dynamic trajectory creating COST that Piaget is trying systematically to describe. It is, yes, but a weak "redescription" of Piaget. (Piaget was marginalized via an accusation that he was merely "redescribing" development in his own language.[30]) The simple activities Thelen et al. note, when viewed as we have above (even if only minimally) from a larger perspective, are part of the emergence of causality, of sequence in time, of the self as object among other objects, of the body as a force among other forces, i.e., an emerging complex of base concepts.

Despite this weakness in accounting for the all-important *h* parameter which drives their whole mass-spring model, Thelen et al. proposed that this general dynamical model *will ultimately apply to all Piagetian tasks*! This would include the stick seriation task and the ordering of two liquid flows discussed above, and many others equally and more complex. This is a severe undervaluing of the problem of representation, i.e., of explicit, conscious thought. If the ordering of two liquid flows example isn't already enough, when we view the Tunnel-Bead task in Chapter VI, we shall see how vastly inadequate is the attempt to reduce Piaget and his children to mass-springs.

Why Cognition Needs Consciousness

Let me summarize the difference between implicit and explicit memory. The implicit relies solely on the fundamental law of redintegration, where a present event (E') redintegrates a past event (E) with a similar invariance structure, or E' -> E. This law operates without a functioning hippocampal-prefrontal system or a functioning dynamic supporting the symbolic function. The law underlies "priming" effects. It is fundamentally mechanical. The explicit, on the other hand, for its development, rests in the development of COST and the symbolic function. It requires the integrity of a very concrete dynamics that participates integrally in the non-differentiable flow of time and thus supports the articulated simultaneity or temporal whole binding the past event with a present event taken as a symbol for the past. Loss of the explicit has nothing to do with the destruction of experience stored in the brain.

In cognitive science, the question has lurked for some time: just what, if anything, does consciousness contribute to cognition? Neither the symbolic manipulation model nor the connectionist net appear to require it or have room for it. For explicit memory, we have seen that a symbolic relationship is required and inherent in this, an articulated simultaneity which implies a global, temporal whole existing over a non-differentiable flow of time. These same principles, following the lead of Cassirer, reside beneath many other cognitive functions – analogy, drawing, number, voluntary action and more.

Let us take perspective. The brain is initially presented an undifferentiated external field (environment). The field is an extensity (not abstract or constructed space, i.e., not a continuum of "positions"), a non-differentiable flow or duration (not abstract time, i.e., an abstract series of "instants"), qualitative (not quantity), a subjective event field (without external "causes"). From this the brain must carve its invariance laws – its "object language." Invariants are isolated by transformations upon this field and the transformations are naturally effected through actions. It is actions that drive the developmental dynamics and this dynamics drives towards a base set of constructs. The base set of fundamental constructs of object, causality, (classic) space and time in which the self takes it place as an object among others, is foundational to the entire structure of thought, which is to say, this set lies at the core of *representation*. Explicit memory, in its developed form, is at its heart a problem of representation – and the brain's achievement of the solution.

Recognizing the Explicit

Explicit memory is clearly considered a more advanced form of memory. Symmetrically, there has been the argument that the implicit "system" is the more primitive – present early in infancy – while the explicit "system" develops gradually, gaining critical mass, as we have seen, around two years. If infants manifest only primitive, implicit memory, a memory by definition "unconscious," then this has even been interpreted by some to mean that infants are not yet *conscious* until this development of the explicit system is complete. This is absurd. As I noted in Chapter II, frogs and chipmunks are conscious. Infants are conscious. Perception *is* consciousness. This latter – that perception *is* consciousness – unbelievably, is not understood, hence the fixation on explicit memory as the key ingredient.[31] Infants certainly have perception. They don't have explicit memory, but they do have a basic form of memory.

Rovee-Collier and associates (developmental theorists) in their extended examination of the entire implicit-explicit distinction in infancy, feel they are able to show that infants display explicit memory way before the completed development of COST.[32] (I think this may have been a reaction to the notion that infants are not conscious because they don't have explicit memory. Therefore, let us show that the explicit exists very early, hence consciousness exists.) These theorists need to have given a little more weight to Piaget. Throughout *The Construction of Reality in the Child*, he makes it clear that a basic form of memory is present; it is just not the developed memory relying on COST which allows for localization of events in time. Thus when he speaks of the form of memory manifesting in the first few stages, he notes that "the only form of memory evidenced… is the memory of recognition in contradistinction to the memory of localization or evocation."[33] Of this recognition, he was not willing to grant it any mature status, noting "… it is not proved that recognition transcends a global sensation of the familiar which does not entail any clear differentiation between past and present but only the qualitative extension of the past into the present."[34]

Rovee-Collier et al., I think it can be said, base much of their argument for the existence of the (very early) explicit in infants on this very form of recognition. For example, an infant is shown a mobile hanging over his crib to which his foot is tied such that he can make the mobile move. At a later time, perhaps a few days, the mobile is shown again, and the infant's memory is judged by his reaction. If he again moves his foot to move the mobile, this is considered a *recognition* of the past situation. Piaget would allow this, but with the qualification already noted above, i.e., this is not explicit memory.

But the status of this *recognition* and its role in memory has been utterly obscured to current theorists by the computer model of mind and/or the information theoretic framework. This is because there is a form of recognition that can be more than just an implicit (unconscious) memory and less than the explicit. It is neither, but is has a status. It can be indeed a *feeling* of familiarity. Such a feeling of course implies a being that exists over a continuous time.

Automatic Recognition

Bergson attempted to distinguish two forms of recognition – "automatic" and "attentive."[35] Attentive recognition, he felt, proceeds "from the subject," while automatic recognition proceeds "from the object." Attentive recognition we have already seen. It resides in the realm of the explicit, and I leave it to the interested reader to explore Bergson more deeply on the subject. The automatic recognition is worth dwelling on briefly for the sake of fleshing out Piaget's concept of early recognition discussed above. Each form of "recognition" appears to correlate with a particular "dissociation." Here are two major dissociations in memory:

1) We have recognition failures, but memory images are intact.
2) We have (a form of) recognition intact, but memory images are lost.

The second is related to explicit memory loss. I am only going to focus on the first here where we have recognition failure but memory images still are available. This state appears to rest on damage to the automatic or "procedural" form of recognition, a recognition proceeding "from the object."

We are all familiar with the experience of moving into a new city and experiencing the initial lack of a feeling of familiarity or recognition with our environment. However, after time, we become quite familiar – our body becomes automatically attuned to the environment. We and our body know exactly where we are, where to turn, where we are going, etc. This feeling, Bergson would argue, arises from the development over time of an automatic motor or action system accompaniment to our perception. The street layouts, gradients, building contours – all bring about an automatic organization of potential actions.

What happens when a lesion to the brain disrupts the mechanisms which create this schema of organized actions? We have the phenomenon of an "aphasic" disorder. We look at the street and no longer recognize it, nor do we have the slightest idea which way to turn. But has the memory of the streets, the city, the normal habitat been lost? As Bergson pointed out, a case described by Wilbrand is pertinent. The patient could describe with her eyes shut the town she lived in, and in imagination walk through its streets, yet once in the street, she felt like a complete stranger; she recognized nothing and could not find her way. What has been lost here, clearly, is not the memory image. What has been lost is a certain action accompaniment or organization built over time relative to the perceived spatial layout. This "feeling" of an action accompaniment, similar to my note on the "felt dissonance" that would arise when observing little geysers while stirring our coffee, should not be ignored as an explanatory and epistemological construct. It is, in current parlance, an *intrinsic* intentionality. This is necessarily a time-extended phenomenon; it is one of the differences between this approach and a disembodied, information-theoretic framework.

Similarly, we are surrounded by objects of our every day experience which are normally associated with certain sets of actions. There are spoons, forks, tables, chairs, bowls, hammers, etc., all of which we have used repeatedly over time. Eventually these are part of action "syntagms." A syntagm is supported by a neural organization where we have actions initially, in infancy, overtly acted out in the presence of such an object – I see a rattle and start shaking my tiny fist. Over the course of development, the overt action is eventually suppressed. Again, what occurs when a lesion to the brain disrupts the mechanisms which organize the normal motor responses to such objects? There occurs an "agnosia" which describes the failure to recognize objects. Thus a victim who was once a writer may be shown a pencil and have not the slightest recognition of its function. Yet the patient can summon a mental picture of the object named and describe it very well.

Moving into a new city has its analogue in learning a new language. A new foreign language presents to us a confused jumble of sounds in which we are unable to distinguish any parts of speech – words, or sentences. Again, over time, an automatic organization of responses is constructed. There is a kind of automatic analysis by the brain of the source and method of production of the speech stream one is listening to. Without getting too deeply into the theory of speech recognition, I think it can be said that at least in this aspect, what is termed the motor theory of speech perception has a great deal of support.[36] This automatic analysis provides a partial sketch or breakdown of what is truly a continuous flow of sound into perceived components. Lesions in the areas of the brain responsible for this analysis cause what is termed "word deafness." The patient is sent back to his original state, unable to distinguish parts of speech in the continuous flow of a speech stream. This hardly needs to be explained in terms of the loss of the auditory images of words stored in the brain (there are no such images). Such patients may still in fact retain the ability to read.

This "automatic motor accompaniment," as Bergson termed it, has multiple components or processing phases, some of which are related to perceptual analysis, some to the grouping of relations, and some ultimately to the relation of perception to action. The case of C.K. is interesting in this regard.[37] C.K. received a severe trauma to the brain in an auto accident. He is very poor at recognizing visually presented objects, yet can name and recognize the same object presented to the touch. His visual recognition failures appear to stem from an inability to segment and group elements in array. For example, he cannot match or trace items that are overlapping and cannot segment a figure from the ground as visual noise is increased. He can pick up local features of an object but not organize them as a whole – thus a dart is called a feather duster, a tennis racket is called a fencer's mask. He copies geometric forms in a slavish, piecemeal fashion without appreciating the identity of the object being copied. Yet despite this recognition deficit his mental imagery, and thus his memory images of objects, is intact. He can draw wonderfully in detail from memory, for example the very objects he cannot recognize, can describe in detail the size, color and shape of letters and objects he fails to identify visually, and can generate novel images.

This automatic recognition is a foundation. In a healthy brain, it would be conceived as evoking a modulation pattern over the brain, a wave supporting redintegration. Whether this state supports localization of a previous event in time, whether it goes beyond Piaget's mere "qualitative extension of the past into the present," depends, from the perspective developed earlier, on the further presence of the greater dynamics supporting COST.

The conventional wisdom ignores this automatic motor accompaniment. The wisdom is that what is here termed automatic recognition relies on the "matching" of the perception to a canonical memory image representation, e.g., we recognize a fork by matching the fork to a canonical or invariant form/image of the generic "FORK." In C.K.'s case, the recognition deficit is explained by the failure of the analytical phase of perception which segments forms and organizes relations. Since the output from this phase is bad, it cannot be related to the canonical form or memory image. Higher order processes, involved in C.K.'s retrieving memories and manipulating them in imagery (i.e., from the "side of the subject"), remain intact, while imagery itself is not assumed to require the perceptual analysis of which C.K. is incapable, i.e., images do not have to be "recognized." The truth is, there is no idea how or where the brain stores the "image" of a fork, or what a canonical fork would be or how it would be "matched." I could not begin to go into the difficulties here on this notion. In Bergson's framework (and I believe in Piaget's) conventional wisdom is wrong. Bergson would say, rather, that in C.K.'s case, the early analytic failure is such that the perception has lost its link to its accompaniment (the action syntagm) in organized action systems (and the feeling of this accompaniment), not that this recognition fails from inability to invoke and match a memory image.

In my opinion, it is in this area – the description of the neural networks that support the "automatic motor accompaniment" – that connectionism finds at least one of its proper homes. Even here, the models will look quite different than the current, "static event" versions. Even walking down my driveway every day towards the barn, I have a flow field, i.e., a *flowing* field. It is for this, with its dynamically transforming texture gradients and constant ratios for size, that a very dynamic neural organization is forming. It cannot simply be a static, "fire once and get a response," network.

Don't Forget Amnesia

If we see damage to the dynamics supporting symbolic relationships as the underlying cause of amnesia, the implication is that this is a general deficit that should affect the memory of *all* events of our past experience. That is, amnesia would have a *unitary* cause. But amnesia presents some apparent difficulties here. There is first the widely accepted phenomenon of limited retrograde amnesia (RA), where for example, memories of two years immediately previous to the damage are lost, but earlier memories are retained. There is, secondly, the phenomenon of "graded" RA. In this case, if we measure decade by decade from the point of the trauma, there is an increase in the number of events remembered the further back in time we move. These two phenomena have long supported the notion that amnesia is a *storage* (in the brain) problem, and led to the long standing notion of *consolidation*, where memories are "dug in" or engrained as it were for a short period of time after the experienced event. The damage to the brain suffered by the amnesic is seen to disrupt this consolidation process.

As a difficulty for the notion that it is destruction to these consolidated memories that is responsible for amnesia, cases are reported in which memories apparently return gradually in the reverse order – oldest to youngest. Further, consolidation is supposed to be a limited process, driven by the hippocampus, occurring over a short period of time. Yet, in an embarrassment to the theory, findings indicate that this consolidation process would have to be happening for *years*, for all events ever experienced, with no limit.

The existence of temporally limited RA (for example, the loss of the just previous two years of events) has become increasingly questioned. Warrington and Weiskrantz suggested that the duration of RA may be underestimated, that memory deficits may extend throughout the amnesic's entire existence.[38] Citing findings of Warrington and Sanders that remote memories in older subjects (age 70-79) were not selectively preserved, they argued for a unitary functional disorder which could account for both anterograde (no memories retained of events experienced *after* the damage) and retrograde effects.[39] Based on his 1997 work, Weiskrantz yet held this position – any memory task that requires the *present x past* product is seen as being at risk.[40]

There are growing numbers of difficulties and with the entire graded RA notion (where we see increasing numbers of events retained as we go deeper into one's past). Nadel and Moskovitch ultimately proposed a "multiple trace" model which sees repeated events creating redundant "traces" within the whole hippocampal complex.[41] Older memories, either being relived mentally more often, or simply repeated more often in the sense of my going to a certain workplace everyday for years, are more resistant to damage because they are more distributed throughout the brain (hippocampus). This, however, simply takes us back to our earlier example of the multiple piano practice sessions of Chopin's Waltz in C# minor. Certainly some set and form of neural traces are being laid down in the brain, both for the resultant motor program to unleash the waltz from memory and for an attunement to the overall environment – the layout of the piano keyboard, the practice room, the furniture, etc. This is not the record of each separate practice experience, which also exists, but it is the neural residue. The more of these traces laid down – the more redundant – the more likely that some survive damage. It is a certainty that such traces can only help to support the fundamental operation of redintegration of a past event with which explicit memory can begin. Simultaneously, the hippocampus may well be a critical component, along with the pre-frontal areas, in supporting the overall temporal dynamic or "temporal whole" required for the symbolic (either of which, to include their connective linkage, could be damaged with various phenomenal results). The greater the damage to the hippocampus (or temporal lobe) the more degraded is this overall dynamic supporting the explicit.

I have examined this broad and complex issue more extensively elsewhere.[42] In general, research and theory has been so dominated by the view that we are looking at damage to memories stored in the brain that few, if any, have asked what the interpretations of amnestic phenomena might be under the opposite hypothesis. I have not penetrated fully into the resolution of all the phenomena in this amnestic area, but I am convinced that research from a different point of view, i.e., that the brain is not the suitcase for memories, will resolve in favor of the general conception advocated here.

What Connectionism Forgets

It is a given that connectionist modelers have weighed in on the problem of amnesia. The focus of their models has been on the hippocampus and the role it supposedly plays in storing memories, or when damaged, not storing them. The models rely on, yes, consolidation processes to cement in these memories.

As a structure, the hippocampal area receives input from nearly all regions of the cerebral cortex, and then sends return loops back out to the original sending area (Figure 4.5). It is tempting to visualize this structure as a central station into which trains of "present" information flow, then to be directed into the great past-holding storage areas of the brain (none of which have ever been found). In this view, destroying this structure stops the flow of events into long-term storage. This temptation should have been avoided.

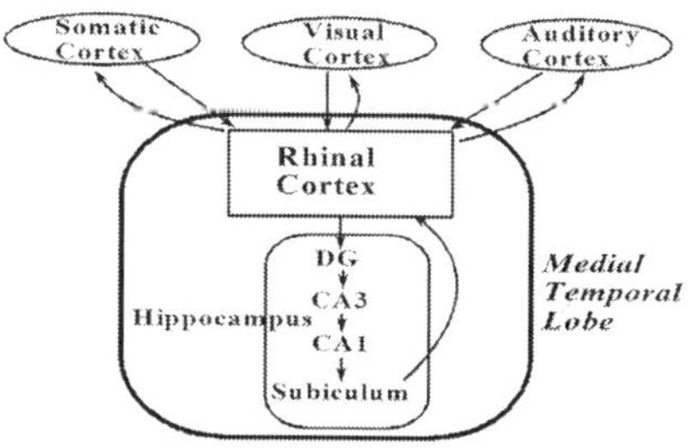

Figure 4.5 The Medial Temporal Lobe System. The neo-cortex regions communicate with the rhinal cortex, and in turn with the hippocampus. This includes the dentate gyrus (DG), CA3, CA1 and subiculum. Feedback loops from the subiculum return to the rhinal cortex and back to the neo-cortex.

Connectionist theorists, O'Reilly and Rudy, outline the conceptual tradition of which their neural network model of the hippocampus is the heir, descending originally from Milner and her colleagues and the famous H.M.[43] The early accounts of hippocampal function emphasized the idea that the hippocampus encodes *stimulus conjunctions*. Wickelgren would describe this as "chunking," a learning process whereby a set of nodes – each node representing an element or, yes, "feature" of an event – are linked to a new, higher order node representing the whole.[44] (Yes, not far from the "geon" approach.) Teyler and Discenna would develop an "indexing" theory, where each experiential event is represented by unique array of neocortical modules (for the visual aspect of the event, for the auditory aspect, etc.).[45] The event, via information flow from the cortex, establishes an index in the hippocampus. This index can be activated by some subset of cues that were included in the original experience, and will activate the entire array of cortical modules for recall of the event. O'Reilly and Rudy also note that Squire, after reviewing the issue, would state his view on how the hippocampus supports declarative (explicit) memory as such:

> In the present account the possibility of later retrieval is provided by the hippocampal system because it has bound together the relevant cortical sites. *A partial cue that is later processed through the hippocampus is able to reconstruct all of the sites and thereby accomplish retrieval of the whole memory.*[46]

Squire is obviously describing redintegration. As usual, there is nothing in Squire's description to explain why this redintegration is *explicit,* i.e., conscious. We see too that O'Reilly and Rudy equally wish to explain redintegration. It is "the conjunction of stimulus features" that is critical to this "conjunctive" model and it is the activation of this conjunction that allows memories to be recalled. The overall architecture has a complementary division of labor. It will assign to the cortex certain kinds of tasks that require very slow learning – the "extraction of generalities" (invariants). To the hippocampus is assigned the *rapid* learning of the *specifics* that define a particular experience, i.e., the conjunctions of features.

We need to go no deeper into the model. The guiding vision here is that "events" are static. We are storing static scenes – a kitchen scene, a snapshot of the rotating cube – with their static "feature" conjunctions. When looking at the diagram of Figure 4.5, we need only to bring into mind coffee stirring, buttering a piece of toast, baseball batters watching a pitcher delivering a pitch, etc. That is, we bring to mind a dynamic view of the time-extended events the brain is processing (or resonating to) to obtain a far different vision. Invariants defined over time are not simply "bits" of information that can flow over circuits into a central processor and then be stuck in storage areas. Invariants cannot be found in an " instant," let alone instant by instant, point by point, travelling along the nerves. The brain, with its many processing areas is virtually "humming" with feedback while such events occur over time. The hippocampal area, if anything, is a dynamic participant in this global, time-extended pattern for the event. There are no "instants" in this dynamics to parse out, to allow things to stop for a moment for the storage of snapshots along the way, i.e., during the time-extension of the event. The hippocampal area and the re-entrant loops it provides could be seen as modulating the *present* state of the brain, and is perhaps more appropriately pursued in terms of at least partially supporting COST or an approximation to it. In other words, the hippocampus may well not have "time" for storage; it is fully occupied with supporting the ongoing event.

The debate over the precise role of the hippocampus has been voluminous, and I have no precise solution, only that this more dynamic context must become the framework in which it is considered.

Between the two visions, there is a point of compromise. Connectionism is concerned with the linking or associating of *arbitrary* event components – creating the elementary "item-bond." The nonsense syllable paradigm of Ebbinghaus with its QEX-WUJ pairs is paradigmatic. These are arbitrary event components and, as I noted, ultimately mere syntax rules for concatenating objects. McClelland et al. (a precursor to the O'Reilly and Rudy model) felt that paired associate learning of the A-B, A-C variety is the exemplar case of amnestic problems.[47] If you remember, the A-B, A-C case looked like this:

List 1 (A-B)	**List 2 (A-C)**
SPOON-COFFEE	SPOON-BATTER
KNIFE-SOAP	KNIFE-DOUGH
BOTTLE-THIMBLE	BOTTLE-PAN

Amnesics have tremendous difficulty with this. The fact is, it is yet another case of the Past x Present product problem of explicit memory, for the poor amnesic (as does the normal person as well), when being tested, must remember that he is being tested on List 1 or tested on List 2. He therefore must work to somehow keep the general event of learning List 1 (the "context") separate from that of learning List 2, and what pairs were in each list/event straight. No connectionist model, at least as currently conceived, will ever model the dynamic (underlying the explicit) necessary to account for this. Connectionism, however, begins with the fact that it sees the pairs in these lists as no different than the QEX-WUJ pairs. They are just arbitrary "patterns" that have to be associated. For a neural network, the great problem, known as "catastrophic interference," is that after List 1 is learned, then when List 2 comes along, the connections/weights for List 1 are destroyed. McClelland et al. attempted to address this.

We have seen that the pairs in the lists above can be treated as concrete events – the knife cutting the soap, the spoon stirring the batter – where the "components" are naturally, intrinsically "associated" via the dynamics of an event. We can term this a "semantic association." I can make this prediction: amnesics will perform remarkably well in this A-B, A-C paradigm if, as in the SPT model (Subject Performed Tasks), they concretely perform the tasks and are given the concrete tests, with precise dynamic cues that I suggested in

Chapter III. Connectionism, we have seen, has no possible place in its (static) theory for these "semantic associations" with their invariance laws. But this is not to say that the case of arbitrary event components does not have its place.

A list of QEX-WUJ-type nonsense syllables is a legitimate case of a series of events with arbitrary components. Similarly, we could argue that the very first time that I ever walked down the driveway to my barn, *everything* was arbitrary in this event – each tree along the way, each rock, each window in the barn. As I do this driveway walk every day (and in doing so, creating my "automatic motor accompaniment"), is the hippocampus struggling to laboriously link every possible "component?" Or is this simply a global, transforming pattern in the brain whose neural connections are, in total, strengthened by repetition? I can arrange a similar arbitrariness in the concrete SPOON-COFFEE and SPOON-BATTER pairs, and in all the pairs in the A-B, A-C list above. By the coffee cup in which the subject is stirring, there is always placed a small cube. By the bowl with cake batter, there is placed a small ball, and so on with some such object for each A-C, A-B pair. I now have the same arbitrariness – an arbitrary component in each event. In testing, I ask the blindfolded subject to stir a spoon in a coffee-like liquid or in a batter-like substance and now ask, "What was beside the thing you are stirring?" (Answer: Cube or Ball.) With connectionism, I can now ask how such an (arbitrary) "association" is formed.

The problem is that when the subject starts stirring the cup, we can imagine the event's invariance structure again forming its dynamic pattern within the brain, just as when walking down the driveway to the barn. Each time the little cube is seen/experienced next to the cup, it is enfolded as part of this event pattern in the brain. This is the question: Must the brain be laboriously "linking" the little cube into this neural dynamic pattern? Or is this (little cube) again just becoming part of the total neural pattern for the perceived event, strengthened by repetition? Or, when I start stirring, therefore re-instantiating a portion of this pattern, is the event/image with the cube being redintegrated – as part of the stack of events (as in Figure 3.12)? I think there is a bit of both the latter two cases, but this, I believe, is the better framework for future research on neural networks.

As I warned, the relation between that which is stored in the brain, and that which is not, is complicated. Unfortunately, we have yet to explore it.

Chapter IV: End Notes and References

1. Cassirer, E. (1929/1957). *The Philosophy of Symbolic Forms, Vol. 3: The Phenomenology of Knowledge.* New Haven: Yale University Press, p. 215.
2. Cassirer, op. cit., p. 184.
3. Winograd, T. (1971). Understanding natural language. In Bobrow, D., & Collins, A. (Eds.), *Representation and Understanding*, New York: Academic Press.
4. Winograd, T., & Flores, F. (1987). *Understanding Computers and Cognition.* Norwood, N.J.: Addison-Wesley, pp. 73-74.
5. Kiverstein, J. (2007). A robot's subjective point of view. Journal of Consciousness Studies, 14, 127-139.
6. Gray, J. A. (1995). The contents of consciousness: A neuropsychological conjecture. *Behavioral and Brain Sciences, 18*, 659-722.
7. Weiskrantz, L. (1997). *Consciousness Lost and Found.* New York: Oxford.
8. Robbins, S. E. (2009). The COST of explicit memory. *Phenomenology and the Cognitive Sciences, 8*, 33-66.
9. Rakison, D. (2007). Is consciousness in its infancy in infancy? *Journal of Consciousness Studies, 14*, 66-89.
10. The first chapter of Piaget's *Insights and Illusions in Philosophy* (1971) can be consulted for some insight into the effect of Bergson on Piaget. This book however goes on to use Bergson virtually as a whipping boy for Piaget's contempt of the "reflective" approach of philosophy as opposed to the "scientific" approach founded in observable facts. Given Piaget's comments on some of the philosophers who appear to have been in his environment, he appears to have had cause. Bergson, however, ever the clear exponent of "tying philosophy to the fact," would have strongly supported Piaget and his work. But Piaget devotes a chapter in this book to the foundering of philosophers upon the rock of relativity theory (and therefore on the "facts"), using Bergson's critique of the theory as the exemplar. But more precisely, Bergson's critique was of the theory's *interpretation* (originally by Langevin) in the context of the twin paradox, and this problem is far from settled. Piaget however makes it quite clear to me that he did not at all understand the issues in Bergson's relativity debate (cf. my Chapter VIII).

    What is ironic, however, is that when the 1971 book was written, Piaget was unaware of the information processing avalanche that was poised to rush over his own intellectual efforts. Indeed, after discrediting a number of his findings, the entire information processing *philosophy* would be lowered over his observations. And if there will be a way to dig him out, it will, in my opinion, be through the fantastic (surely to Piaget) Bergsonian model of the brain as a modulated reconstructive wave within a holographic field, and the definition of the role of Piaget's developmental theory of *operations* (see my Chapter VI) *in this context.*
11. For a discussion of this development in more detail, see Robbins, 2009, op. cit.
12. Van deer Maas, H.L.J, & Molenaar, P. C. M. (1992). Stagewise cognitive development: an application of catastrophe theory. *Psychological Review*, 99, 395-417.
13. Piaget, J. (1954). *The Construction of Reality in the Child.* New York: Ballentine, p. 103.
14. Piaget, J. (1952). *The Origins of Intelligence in Children.* New York: International universities press, p. 391.

15. Cassirer, E. (1929/1957). *The Philosophy of Symbolic Forms, Vol. 3: The Phenomenology of Knowledge.* New Haven: Yale University Press.
16. Cassirer, E. (1929/1957). *The Philosophy of Symbolic Forms, Vol. 3: The Phenomenology of Knowledge.* New Haven: Yale University Press, p. 236.
17. Head, H. (1926). *Aphasia and Kindred Disorders of Speech (Vols. I and II).* Cambridge: Cambridge University Press.
18. Cassirer, op. cit., p. 250.
19. Cassirer, op. cit., p. 250, emphasis added.
20. Cassirer, op. cit., p. 271, emphasis added.
21. Cassirer, op. cit., p. 257, emphasis added.
22. In one of the cases above, Cassirer discussed ideomotor apraxia, often defined as "inability to carry out a motor command, e.g., act as if you are brushing your teeth." It is a phenomenon for which lesions to an area in the brain termed the "dorsal stream"are currently implicated, or perhaps the ventro-dorsal as Gallese recently argues. Gallese, in his review, argues that different and parallel parieto-premotor networks which receive visual information processed within one part of the dorsal stream create internal representations of actions, to include action preparation, action understanding, space and action conscious awareness. But again, the simple notion of an "information stream" (or its disruption) is insufficient to explain the specific problem Cassirer is discussing in relation to action, or for that matter, in the problems relating to sketching, analogy, number and time which all appear to share the same cause. We are asking how "the unities can be fixated, yet simultaneously be kept mobile," or how one and the same element can oscillate between one meaning and another, while comprising a whole. This bespeaks of a dynamic, in the very concrete sense I have indicated, which is intrinsically participating in the concrete, indivisible or non-differentiable motion of the matter-field in time.

    Gallese, V. (2007, April). The "conscious" dorsal stream: embodied simulation and its role in space and action conscious awareness. *Psyche*, 13 (1), 1-20.
23. Piaget, J. (1968). *On the Development of Memory and Identity.* Clark University Press.
24. Piaget, J. (1927/1969). *The Child's Conception of Time*. New York: Basic Books.
25. Rogers, T., & McClelland, J. (2004). *Semantic Cognition: A Parallel Distributed Processing Approach.* Cambridge: MIT Press.
26. Quinlan, P., van der Maas, H., Jansen, B., Booij, O., & Rendell, M., Re-thinking stages of cognitive development: An appraisal of connectionist models of the balance scale task. *Cognition*, 103, 413-459.
27. Markovitch, S. & Zelaso, P. (1999). The A-not-B error: Results from a logistic meta-analysis. *Child Development, 70*, 1297-1313.

    Munakata, Y. (1998). Infant preserverative and implications for object permanence theories: A PDP model of the A-not-B task. *Developmental Science, 1*, 161-184.
28. Thelen, E., Schoner, G., Scheler, C., & Smith, L. (2001). The dynamics of embodiment: A field theory of infant preseverative reaching. *Behavioral and Brain Sciences, 24*, 1-86.
29. Ibid., p. 31, emphasis added.
30. Brainerd, C. J. (1978). The stage acquisition in cognitive developmental theory. *Behavioral Brain Sciences*, 2, 173-213.

31. My first published paper, in the *Journal of Consciousenss Studies* (2000) - "Bergson, Perception and Gibson" - was nearly rejected by a reviewer because he thought it was "only about perception," not the problem of consciousness.
32. Rovee-Collier, C., Hayne, H., Colombo, M. (2000). *The Development of Implicit and Explicit Memory*. Amsterdam: John Benjamins.
33. Piaget, J. (1954), op. cit, p. 369.
34. Piaget, J. (1954), op. cit, p. 369
35. Bergson, H. (1896/1912), op. cit.
36. Liberman, A. M. & Mattingly, I. G. (1985). The motor theory of speech perception revised. *Cognition*, *21*, 1-36.
37. Behrmann, M., Moscovitch, M., Winocur, G. (1994). Intact visual imagery and imparied visual perception in a patient with visual agnosia. *Journal of Experimental Psychology: Human Perception and Performance*. *20*, 5, 1068-1087.
38. Warrington, E. K., & Weiskrantz, L. (1973). An analysis of short-term and long-term memory defects in man. In J. A. Deutsch (Ed.), *The Physiological Basis of Memory*. New York: Academic Press.
39. Warrington, E. K., & Sanders, H. I. (1971). The fate of old memories. *Quarterly Journal of Experimental Psychology*, *23*, 432-442.
40. Weiskrantz, L. (1997). *Consciousness Lost and Found*. New York: Oxford.
41. Nadel, L., & Moscovitch, N. (1997). Memory consolidation, retrograde amnesia and the hippocampal complex. *Current Opinions in Neurobiology*, 7, 217-227.
42. Robbins, S.E. (2009), op. cit.
43. O'Reilly, R. & Rudy, J. (2001). Conjunctive representations in learning and memory: principles of cortical and hippocampal function. *Psychological Review*, 108 (2), 311-345.
44. Wickelgren, W. A. (1979). Chunking and consolidation: A theoretical synthesis of semantic networks, configuring in conditioning, S-R versus cognitive learning, normal forgetting, the amnesic syndrome, and the hippocampal arousal system. *Psychological Review*, *86*, 44-60.
45. Teyler, T. J., & Discenna, P. (1986). The hippocampal memory indexing theory. *Behavioural Neuroscience*, *100*, 147-154.
46. Squire, L. R. (1992). Memory and the hippocampus: A synthesis from findings with rats, monkeys and humans. *Psychological Review*, *99*, 195-231. (Squire, 1992, p. 224, quoted by O'Reilly & Rudy, 2001, p. 313, emphasis added)
47. McClelland, J.L., McNaughton, B.L., & O'Reilly, R.C. (1995). Why there are complementary learning systems in the hippocampus and neocortex: Insights from the successes and failures of connectionist models of learning and memory. *Psychological Review*, *103* (3), 419-457.

CHAPTER V

The Koan of Action: Free Will

Yes, we have a soul, but it's made of lots of tiny robots.

- Daniel Dennett, *Freedom Evolves*

The least act, such as eating or scratching an arm, is not at all simple. It is merely a visible moment in a network of causes and effects reaching forward into Unknowingness and back into an infinity of Silence, where individual consciousness cannot even enter.

- D. K., Canadian housewife, in *The Three Pillars of Zen*

Freedom From Robots

I have taken us progressively through the counter-conception of mind, far, far from the realm of the Terminators, Datas, R2D2s, Spielberg's AI-child, and other "intelligent" machines that populate the mythical world of today's underlying theory of mind. The concept that we are not free is thoroughly embedded in the robotic, machine conception of being. The idea of determinism, determined action and rule-driven machines, be they computers or robots, stems from the same primitive conception of space and time from which the whole sorry machine conception takes its origin. Bergson utterly destroyed it long ago. He did so in his doctoral thesis and subsequent book in 1889, entitled, *Time and Free Will.* This was where he first expressed his profound insight into the nature of time. His entire philosophy was born in the moment of his absorption in the *concrete* experience of time.

Figure 5.1. Henri Bergson

This was not the Henri Bergson of a slightly earlier frame of mind. He had been absorbed in the philosophy of Herbert Spencer. Spencer was then the foremost exponent of the great mechanistic view of the universe and physics of the time, summed up in the vision of the physicist, Pierre Laplace (1749-1827). Laplace held that given sufficient knowledge of the states of the particles of the universe at an instant, one could, knowing the right equations or laws, completely predict any future state of the entire universe.[1] Bergson's devotion to Spencer and his mechanism was well known to his fellow students, and produced a scene in the university library. A professor, seeing some books on the library floor, turned to Bergson (then student librarian), and said, "Monsieur Bergson, you see those books sweeping up the dirt; your librarian's soul ought to be unable to endure it!" To which his classmates exclaimed, "But he has no soul!"[2] Yes, Dennett would have loved him – then.

But his study of Spencer's concept of time was to change this. To Spencer, time was to be treated as if it were little different from space. Like space, time is measurable and contains juxtaposed parts. Like space it is a homogeneous medium whose parts and properties are everywhere alike. Like space then, it is capable of being measured by mathematical concepts, and one can say that one time is equal to another, or twice as long as another, etc. And finally, it is meaningful to say that time is composed of instants and that the movement of time can be viewed as a series of instants, as a body moves from point to point in space.

To his surprise, Bergson found that this common sense and well accepted view held little validity. The most cursory examination revealed that moments of time are not alike, that each has its own quality, nor do instants ever simply repeat, or simply occur in succession. Rather there is a continuous flow where each moment merged into the next. And thus, as we have seen, Bergson was to describe a far different time as lived by the self, a time he termed *duration*.

> Below homogeneous [abstract] time, which is the [spatial] symbol of true duration, a close psychological analysis distinguishes a duration whose heterogeneous moments permeate one another; below the numerical multiplicity of conscious states, a self in which succeeding each other means melting into one another and forming an organic whole.[3]

Physical time, as described by Spencer, could only be the limiting case of this *real* time. But such a "time" entailed the rejection of Spencer's mechanism. If mechanism was correct, all change is reducible to the completely predictable motions of material particles, and "free will" is an illusion. But the experience of duration showed that change was far more dynamic, organic, and interrelated as a process than could be supported by the motions of particles. The present could not repeat the past; each present moment was poised before novel possibilities. Mechanism was wrong both in its account of change and in its assumption of simple prediction.

In 1889, Bergson defended and then published his doctoral thesis, *Time and Free Will*, which held his analysis of the problem of time in the context of the then newly born science of psychophysics, and contained as well many of his insights into the physical world. To his own chagrin, his advisor, the psychologist Janet, thought only the first chapter with its analysis of sensation was significant, while the philosopher Boutroux commented only on an aspect of the thesis concerning its concept of liberty. Though furious, Henri's good manners prevailed. It is an unfortunate fact that the same misjudgment of the significance of this work resumed with the decline of Bergson's popularity. The perennial philosophical discussions on free will and determinism carry on today, sadly, without the smallest of references to Bergson. It is as though all the participants are literally hypnotized by the illusory reality of abstract space.

The problem of the existence of free will collapses before the understanding of the nature of the flow of time. I am not saying there are not still mysteries, but the debate on whether mind is free becomes absurd. The philosophers of the machine conception of course think otherwise, and we shall look at their foremost modern exponent shortly, but before we do, let us consider another remarkable aspect of our ability to act, one that leads to one of these mysteries.

Monkey Not Do, Monkey Not See

The principle of virtual action, with its intrinsic relativity, envisions a level of integration of perception and action beyond any theory today. No robot, as currently conceived, can implement this relationship. In the context of the Turing Test therefore, expanding the nature and method of the Test only slightly, we should be able to say to our robot, "Sir, please take this little pill (containing various pertinent catalysts) which will change your capability of action. Now please describe your perception." The robot should begin speaking of heron-like flies, or, were the robot initially "perceiving" a cube rotating with sufficient velocity such that it is a fuzzy cylinder (a figure of infinite symmetry), he now should begin speaking of a serrated-edged figure of 4n-fold symmetry (for he must continue to perceive by the same invariance laws specifying the event). Should we say, "Sir, please ingest a little more catalyst," he should begin speaking of a cube (a figure of 4-fold symmetry) in slow rotation, and should we ask him to ingest a bit more, he begins speaking of a stable, motionless cube. And should we ask him to now reach for the cube, he modulates his grasping apparatus to grasp stable edges and sides, as opposed to a cylindrical figure. That is, this is what we should expect of a being for whom perception is virtual action.

Currently, in Cognitive Science, there is discussion of the "perception-action cycle." In this cycle, information is received from the external world by the vision systems and sent on to systems for action. Action then occurs which modifies the world, creating a new perception, with new information for the action systems, and so on. Yes, it is in essence the "comparator" model of the previous chapter. There is little grasp of the implications of the motor systems actually feeding back their results to the visual areas, determining how we view the world (as action possibilities). Virtual action naturally makes a prediction for this event: Sever the neural tracks connecting the visual areas in the cortex to the motor areas. What is the expected result? If vision is the display of possible action, and there is no information from the action systems, then there should be no vision. One should be blind.

This is not a currently expected result. Theorists such as Francis Crick (*The Astounding Hypothesis*), for example, argued that consciousness (that is, conscious visual perception) is completely a function of the connections of the thalamus with areas in the cortex. This hypothesis has been superseded, but the other current theories equally ignore the motor areas. Yet there is already a disturbing result that is ignored, disturbing unless you are aware of the virtual action hypothesis. The neuroscientists, Nakamura and Mishkin, already asked the "sever the connecting tracks" question in the early 1980's. They indeed severed the tracks between the visual areas and the motor areas in monkeys. The monkeys became utterly unresponsive to external events. They were for all practical purposes blind; in fact, as the theorist of memory and consciousness, Lawrence Weiskrantz, argued, they were blind.[4]

Mr. Ted Williams

As I realized these implications of Bergson, while scribbling away on my Ph.D. thesis in the early 70's, memories flashed of a favorite topic of my dad when I was a child. Dad loved Ted Williams. Of course Dad was from Massachusetts and Williams played baseball for the Boston Red Sox all his life. He was arguably the greatest hitter in baseball. He is the last hitter to hit over .400 (.406 in 1941), a mark not reached again for over sixty years now and counting – despite the interim advent of steroids. He also hit in the .400's in 1952 and 1953.

Dad loved to tell me how Williams could read the label on a record as it spun around on the phonograph turntable. He could also, it was reported, actually see the seams and the spin on the baseball as it hurtled towards him. Dad would tell me how Williams could "wait longer" than other players before he initiated his swing.

What does this sound like? It sounds, for all intents and purposes, as though Williams was in our "higher energy state." Watching a fly, he would have seen it moving more slowly – precisely because he could act more quickly. He "could wait longer" to swing because he was seeing, in the ball's slower motion, just how he could act, just like our cat watching the mouse.

In my graduate school days, when I talked about this theory occasionally, friends would respond with similar experiences. One told me that when he played basketball, on occasion, when driving to the basket, everything would slow down – the motions of the other players, the ball – and he would weave unimpeded to the basket. On other occasions, in baseball when batting, the ball would slow down and seem very large, and he seemed to have all the time in the world to smash a hit. Similar stories are not uncommon with people in the midst of accidents such as car crashes.

All this begs certain questions. Was this Williams' permanent, physical, "energetic" state? If so, why? Did he have to "slow down" to talk with normal people at normal scale? Or was it a state he could voluntary induce? If so, how? What does this say about the mind-matter relation? In the case of my graduate student friend, it was clearly not permanent. Then what brought it on? Simply an excitement, like an adrenaline rush, something that acts like our "catalysts" – a cause also assigned to this slow-down of events during accidents? Or something else? The questions are far larger and more profound than realized, though if we stay stuck in the sterile computer metaphor of mind, they will never be asked. I will explore these more in the thoughts to come.

On the flip side of increases in underlying chemical velocities, there is the unfortunate problem of aging. The little old lady, driving timidly down the freeway in her car, peering just over her steering wheel, is seeing the other cars whizzing by at blazing speed precisely because the world reflects her decreased ability to act. The research literature on aging (or gerontology) is filled with reaction time and perception experiments that indicate this.[5]

Voluntary Action – the Intent or Atemporal Idea

The problem of voluntary action can be stated simply: how do we will to move our finger? Variants are: how do we will to swing baseball bats, or leap at mice, or reach out and grasp wing-flapping flies? Cognitive science has no clue today. The answer is far from clear, and puts us in the realm of the "interface" between mind and matter.

Of course, in Bergson's framework, we are already given a clue: there is no "interface" between "mind" and "matter." These are not two different "substances," things, or whatever, coming from incommensurate realities such that we will never figure out a means of interaction. We have seen that the matter-field, in its non-differentiable time-evolution, has the elementary attributes of mind – an elementary form of awareness defined throughout, and an elemental, "primary memory" via the indivisibility of this motion. I suspect that it is in exploring the depths and implications of this latter statement that we will develop a truly deep theory of voluntary action.

Our bodily action involves the unfolding of ordered elements in time. These "elements" might be the words of a sentence, or the various muscular movements underlying a dance step or a tennis serve. The ordering or order of the elements is called the *syntax* of the act. Karl Lashley, one of the great theorists of psychology, pointed the field initially into the problem in his 1951 treatise, "The Problem of Serial Order in Behavior."

To pronounce the word "right," he noted, consists in these elements:

1) The retraction and elevation of the tongue, expiration of air and activation of the vocal cords.
2) Depression of the tongue and jaw.
3) Elevation of the tongue to touch the dental ridge, stopping the vocalization and forceful expiration of air with depression of the tongue and jaw.

He noted that these movements have no intrinsic order of association. "Tire" requires these same movements in reverse order (or syntax). The order is imposed by some organization other than direct associative connections. This is equally true for the letters of a word. The letters – "R" or "G" or "A" – can occur in any order or combination. The order depends upon a *set* for a larger unit of action, in this case the word.[6]

Words stand in the same relation to a sentence as letters to the word. Words themselves, he noted, have no *temporal valence*. The word "right" is noun, adjective, adverb and verb, has four spellings and at least ten meanings. In the sentence, "The mill-wright on my right thinks it right that some conventional rite should symbolize the right of every man to write as he pleases," the arrangement of the individual elements or words, the order, is determined by a broad scheme of meaning. This holds not only for language, but for all skilled movements.

What then determines the order, Lashley asked? Lashley took the view that *the set or the idea does not have a temporal order*; that all of its elements are *contemporal*. Thus I read a German sentence, pronouncing the words with no thought of their English equivalents, then proceed to give a free translation in English without remembering a single word of the German text. "Somewhere between the reading and free translation," Lashley notes, "the German sentence is condensed, the word order is reversed, and expanded again into the different temporal order of English."

Lashley argued that the mechanism which determines the serial activation of the motor units is relatively independent, both of the motor units and of the thought structure. Supporting this, he pointed to a series of examples of mistakes of order. He had been tracking typing errors for some time. Misplacing or doubling a letter (t-h-s-e-s for *these*, i-i-l for *ill*, l-o-k-k for *look*), he argued, show the order dissociated from the idea. Further there are "contaminations," often from anticipation, for example typing "wrapid writing," or the Spoonerisms made famous by a probably somewhat aphasic Professor Spooner, e.g., "Let us always remember that waste makes haste." These latter, he argued, indicate that a set of expressive elements seem to be partially readied or activated prior to the overt act.

Thus, Lashley argued that there are at least three sets of things to be accounted for:

1) The activation of the expressive elements (words or adaptive acts)
2) The determining tendency, set or (atemporal) idea
3) The syntax of the act – a generalized pattern or schema of integration which may be imposed on a wide range of specific acts (upon individual components of an act)

In the 1960's, the great linguist, Noam Chomsky, inspired by Lashley, would initiate a vast assault on (3), the theory of syntax, overturning Skinner and his simple "stimulus-response chain" Behaviorism as he did so. He proposed a rule system or grammar, employed by the brain, to unfold the words from the idea in the proper order. In Figure 5.2, the word-elements of the sentence, "The man stirred the coffee" are unfolded in a certain, grammatical order by a set of syntactic rules. A sentence S, for example, by rule, may consist of a *noun phrase* and *verb phrase* (S -> NP + VP), and a noun phrase consists of a determiner (Det) and a noun (N), and so on. The end-elements of the rule structure (nouns, verbs, determiners) come from the pool of possible words or symbols in the language. But to me, there was always a profound implication in Lashley that Chomsky appeared to assume, and the subsequent forty years of cognitive psychology has ignored.

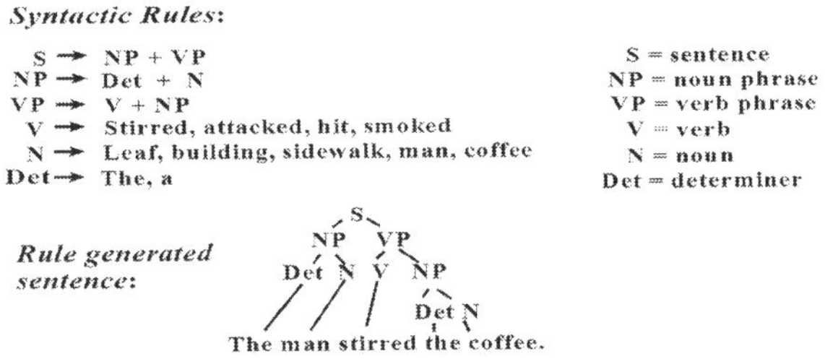

Figure 5.2. A syntactic rule structure, with a "coffee stirring" sentence it generates. The rule, S -> NP+VP can be read as a sentence S "consists of" a NP (noun phrase) + a VP (verb phrase).

At the apex of every diagram, such as Figure 5.2, is "S". What is behind S? Behind S is, of course, the *idea*. It is the starting point from which every sentence derives its ultimate form, its syntactic structure, its embodiment in specific end-elements or words, i.e., components of the sentence as an *action*. But this idea, because it represents a multi-modal event, can be expressed in multiple degrees of complexity, richness and detail. From the sparse, "The man stirred the coffee," the same idea could have been expressed as:

- The man quickly stirred the coffee.
- The old man slowly, painfully stirred the coffee.
- As the man stirred the coffee, the aroma wafted everywhere.
- The man quickly stirred the coffee with the spoon clinking away against the cup.

Cognitive science has pursued this symbols-and-rules approach for over forty years. I cannot imagine Lashley being impressed. To begin with, we have ignored his starting point, the atemporal idea. No post-Chomskian computer translation program, in all our subsequent syntactic sophistication, has ever translated a German sentence into an atemporal idea, then freely into English. No computer simulation program, in fact, has ever had an atemporal idea. We have had no theory of "S," the point from which syntactic structures must devolve in an ordered act. In fact, cognitive science, as DORA for example, has tried vainly to reduce "S" itself to data elements and yet more syntax. Yet it is from "S" that the expressive elements must be activated upon which this syntax operates. There is no theory of this activation given the nature of "S." From this perspective, the attack on the true problem of syntax has not yet begun.

The Atemporal Idea

The dream literature is noted for viewing dreams as atemporal forms. Reports from personal experiences describe entire sequences of events taking place in what is apparently fractions of a second in normal time. Bergson describes a case from one individual:

> I am dreaming of the Terror; I am present at scenes of massacre, I appear before the Revolution Tribunal, I see Robespierre, Marat, Fouquier-Tinville....; I defend myself, I am convicted, condemned to death, driven in the tumbril to the Place de la Revolution; I ascend the scaffold; the executioner lays me on the fatal plank, tilts it forward, the knife

> falls; I feel my head separate from my body, I wake in a state of intense anguish, and I feel on my neck the curtain pole which has suddenly got detached and fallen on my cervical vertebrae, just like the guillotine knife. It had all taken place in an instant, as my mother bore witness....[7]

The "man stirring coffee with the spoon" is inherently a time-extended event, however, as a thought, what is its duration? If my intent is to stir the coffee with the spoon, the elements may be conceived as atemporal, in fact an invariance structure defined over 4-D experience. Mozart once reported that in his creative process, he would see an entire piece of music, "though it be long," even unto a symphony, "as though it were a statue." He was speaking of a time-extended, yet atemporal structure. Indeed, though we speak of 4-D extended events, the 4-D memory of which have been treating and its events have no particular scale or duration. The detailed theory of the unfolding of an atemporal idea is a challenge. Bergson treated some aspects of this in his treatise on *dynamic schemes*.

Imagine learning to dance, he asked. To do so, we begin by watching people dance. The result is a visual impression of the movement. It is not a precise image, for the image has yet to be concretely filled in, articulated so to speak, by the actual physically performed movements that will comprise the ultimate fluid action. These movements are elementary units that will be coordinated – movements used in walking, lifting up on one's toes, swinging the arms, etc. But the first impression is an outline of the relations, especially temporal, but also spatial, of the successive parts of the movement. This abstract image of the spatial-temporal relations is the dynamic scheme.

This abstract scheme, he argued:

> ...must fill itself with all the motor sensations which correspond to the movement being carried out. This it can only do by evoking one by one the ideas of the sensations, or, in the words of Bastian, the "kinaesthetic images," of the partial elementary movements composing the total movement: these memories of motor sensations, to the extent that they are revivified, are converted into actual motor sensations, and consequently into movements actually accomplished.[8]

The above passage of Bergson contains, in a nutshell, the deepest questions facing this model which I have been describing when it comes to the frontier of voluntary action. Firstly, the scheme, as I understand it, is not stored in the brain. It is a creature of 4-D memory. Therefore, we are again coming to grips with the nature of the whole transforming matter-field of which any individual being is a part. Correlatively, it is not for nothing that Bergson's article is entitled "Intellectual Effort." He is speaking of the *effort* needed to transduce an abstract scheme into concrete images, images which will in turn induce movements. Effort is another term for *force*, and this in turn sets us into the problem of *just what force is*, once we probe beyond Newton's definition. Finally, we have the picture of the images integrally wound, causally, into the motor unfolding of the act.

The lack of this conception in current theory, where the atemporal idea unfolds into action and the larger vision of the nature of the "subject" this implies, was the core of a significant controversy, to this day very unresolved. In a famous 1985 study by Libet, subjects were asked to perform an act, like pushing a button, at their will.[9] They were to note the point in time, by watching a special oscilloscope display, when they had the conscious intent to make the act. The studies reveal that there is a significant buildup of neural activity 300 milliseconds *before* the time the subject reports the intent to act. Given that the intent is reported *after* the neural activity buildup, the finding led to an ongoing controversy over the actual role of free will, for it could be argued that the subject has no free will at all and is simply reporting an "intent" that is nothing more than an unconscious, determined action already unrolling. It also led to Libet's suggestions – perceived as controversial – of "backward referral in time" of the intent.[10] But the gradual production of neural processes by an (initially "unconscious") idea, such that it eventually, via this means, becomes conscious, is an integral part of Bergson's model, for it is only by this means – the participation in a state of action/processing in the brain – that an idea *can* become conscious. From this perspective, other than showing that there is much we do not understand about the "unconscious," it is hard to see how Libet's result is so unexpected. Rather, it is exactly what should be expected in a larger, four dimensional view of mind, where the "subject" is a time-extended being.

Time and Voluntary Action

The notion of images driving an action is anathema to cognitive science. This is precisely because the image has neither an explanation of its origin nor a useful place in cognitive science. All is simply data elements and rules. For the neural network arm of cognitive science, there is only strengths of connections between neural network units or nodes. Marc Jeannerod, however, reintroduced the concept of images driving actions in a 1994 article in the prestigious journal *Behavioral and Brain Sciences*.[11] Lacking a theory of the origin of images in the first place, Jeannerod was slightly ambiguous as to their actual role, for if experience is merely stored in the brain, images are simply generated from these stored elements. What role do they serve? They appear redundant (or "epiphenomenal").

Is the image causing the physical actions, or are the physical actions causing the image? For the dreaming cat, flicking his tail and twitching his paws, are the dream images of the mouse driving the physical effects, or are the dream images merely epiphenomenal? Though Jeannerod tended to have it both ways, Bergson was unequivocal. He saw the image as an integral phase of the causal flow starting from the intent to act, to the image, to the unfolding of the concrete actions. Let us look at the larger context of research that might be needed to get a grasp of this flow from the intent to the action.[12]

Jeannerod took particular note of the research literature on the real effects of mental imagery practice on motor skill. These studies require the experimental subjects to practice a certain action in their imagination. After some number of these purely mental practice sessions, they are tested to see if they have improved their speed and/or accuracy at the task. The studies show positive results. But there are studies in this context with more radical implications.

In the early 1950's, two researchers, Linn Cooper and Milton Erickson, published the results of a series of experiments they had performed using hypnosis with imagery practice instructions (*Time Distortion in Hypnosis: An Experimental and Clinical Investigation*).[13] The subjects were taught how to self-induce a trance to "alter time." In this altered-time trance state, they would be asked to simply imagine events, or else to practice actions. For the one experiment they did on motor skill improvements in these practice sessions, greatly condensed in time, their results were marginal. Yet, curious incidents were observed. One subject was mentally watching a movie in this altered time state. He was observed to be making extremely rapid movements with his hand, moving between his mouth and legs. When asked after the session what he was doing, he reported that he had "been eating popcorn." The velocity of his actions had been commensurate with the scale of time in his trance-altered state. Another subject, a violinist, was mentally practicing violin passages in altered time. She reported that in this state, she was able to view the whole piece, as though it were laid out as a static structure. We are reminded of Mozart's seeing an entire symphony as though it were a statue.

In our original discussion of altering the time-scale, and action within this scale, I relied on the physical effect of catalysts. In Cooper and Erickson's studies, we are seeing indications of similar effects from the "side of the subject." This is why the source of a phenomenon such as that witnessed in Ted Williams or my basketball playing graduate student friend is an open question, worthy of serious research. To cognitive science, in its limited concept of the brain, the question means nothing. But the studies are very real.

How could this be possible? In the end, it will come to this question: what is "will," and why is it free?

Pin-balls and Free Will

The problem of free will, and therefore voluntary action, is integrally tied to the mechanical conception of the universe. The penultimate version of this conception was coined by our earlier mentioned physicist, Pierre Laplace. In this, he imagined an intellect, now affectionately known as Laplace's Demon, which has knowledge of the exact positions and velocities of every particle in the universe at the beginning of time. Given the demon knows the laws that act on these particles, it can predict the precise, complete state of the universe at any subsequent time. All is determined. There is no free choice or action. The classic scenes for philosophical debate on free will all take place with this Laplacian background. Daniel Dennett (*Freedom Evolves*) accepts this Laplacian, deterministic vision completely. Nevertheless, he spends many pages attempting to convince us, in essence, that if we move up to the macro-level of experience – the world of golf ball putting and making decisions in the stock market – that things are so complicated that we are *effectively* free. His entire discussion produces one of the great "cognitive dissonance" headaches ever achieved as he feverishly attempts to keep our eyes averted from his Laplacian world, lurking always, driving everything beneath his world of macro-appearances.

Let us paint a classic scene of the debate: Before a certain philosopher there sit two glasses of wine, a Burgundy and a Bordeaux. The Burgundy is tasty, but, hmmm, the Bordeaux… There is a moment of hesitation. Drum roll. The philosopher reaches for the Bordeaux. He has exercised his free choice. Or has he? We diagram the event (Figure 5.3). Our philosopher has spent a life of wine drinking, action and heavy philosophizing before reaching this point of decision. We represent this as the portion of the line from O to D. At D we picture the oscillation of the decision, then the path is taken from D to Bx – the sipping of the Bordeaux. The path from D to By – the Burgundy – is equally open, but is not taken, though it could have been. Or could it?

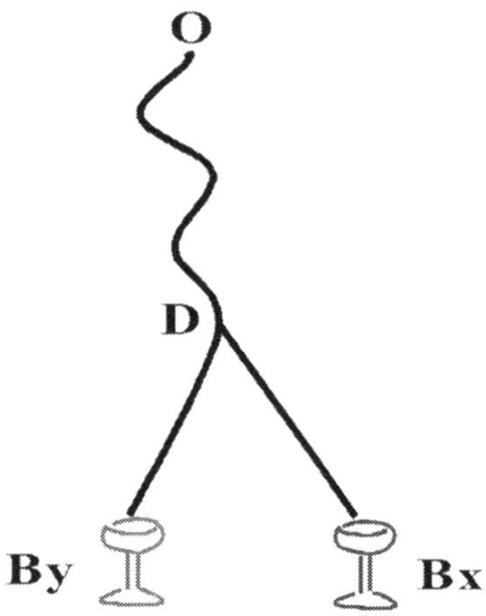

Figure 5.3. The choice – Burgundy vs. Bordeaux

This has been the dilemma. The determinist envisions a form of mechanical oscillation in space at D, almost as a pin-ball at an island-bumper in a pin-ball machine, then states that the antecedent causes are sufficient to account for the path ultimately taken, and indeed must have been in effect to account for the movement. The proponent of free will sees the same oscillation, and says no. Both paths were indeed equally possible. Being unimpressed by the fact that a path was actually taken, he effectively moves the self back to its oscillation at D. In doing so, he ignores Laplace's demon. For the demon, there is only one possible path, and the philosopher's choice is that path.

Bergson (*Time and Free Will*) argued that both are wrong. The natural symbolism of Figure 5.3 by which we represent this decision process treats the motion of the self – yes, yet again – as though it traversed a *trajectory* or line in space. It reduces the progress of the self, i.e., a dynamic motion in concrete, indivisible time, to its residue in space given by a line, the parts of which, like the path of poor Achilles chasing the hare, we can then carve up at will into an infinity of "points" or mutually external "states." This lends itself instantly to a mechanical conception of the self and a mechanical statement of the problem. Each "state" now becomes a "cause" (like the "particle" at a "position" in space) determining the next "state." But this should give us pause. We have already seen the difficulties, in fact the invalidity, in treating motion via the concept of a trajectory in abstract space.

The Projection Frame Again

In truth, we know *nothing but* our conscious perception. All experience is given within in it. All thought is contained within it. All forms of thought derive from it. Our science is built upon concepts *projected* from it. But we have already asked if this layer of projected forms in fact obscures the fundamental structure of the field from which it arose, and now even more so when we use this layer of abstract forms to reflect back upon our own consciousness. It is precisely this layer, we noted, that has been the challenge for physics.

In this projection frame, we saw that the concept of a trajectory begins with the fundamental partition effected by perception within the universal field, dividing it into "objects" and their "motions." Extended in thought, it became the continuum of positions, the completely *relative* motions of "objects" traversing sets of points in this continuum, and the loss of all *real* motion. At stake now, I noted, is a physics which can incorporate real motion. Though we conceive of objects and their motions, Bergson had argued, why can we not conceive of the *whole* as changing, i.e., a global transformation of the universal field, as the turning of a kaleidoscope? The motions of objects become then changes of state – wave motions – within this field. The motion of this whole cannot be conceived as a series of instants or discrete states, but as a non-differentiable, melodic flow.

In this flow, each moment interpenetrates the next, as the notes of a melody, forming an organic continuity. Because each new moment is reflection of the preceding history of the flow, each new moment is absolutely – new. No moment can ever be repeated. But we project backwards upon our conscious experience through our framework of "objects" in an abstract space. We see each note in a melody as a mutually external, repeatable object or state. Ten middle Cs are struck on the piano in succession. We see the spatial cause – the same finger hitting the same key producing the same result. Thus this flow in time becomes divided conceptually into ten mutually external, repeatable "notes." But our actual *experience* is of an organic whole. If one note is held slightly longer than the others, the experience, the quality of the *whole* is changed – the notes interpenetrate and each current note is a reflection of the preceding notes in the series. It is the conscious experience that is telling us the reality, not the abstraction derived from space. The finger is not the "same" finger from note-strike to note-strike. The piano key is not even the same key. The string is not the same string. Nothing is the same in this universe from moment to moment. It is only *practically* the same – a practicality as defined by a scale of time and action calibrated to perception, a scale in turn defined by the dynamics of the brain. To turn this practical, this "middle" scale and the concepts derived there from into philosophical gospel – this is the problem. It makes all the difference when we are trying to get beyond "customary images relative to our needs" not only in physics but in a phenomenon so profound as free will.

But this framework gave birth to causality as well.

Mechanical Causality and Repeatability

The essence of the concept of causality in the classical framework is *repeatability*. Causality, as an explanatory concept in science, is useless if the cause is not repeatable. If I line up a cue ball in the same precise position relative to the 8-ball, apply precisely the same vector of force via the cue stick, I expect the same result. My prediction scheme via Newtonian laws works wonderfully. If I am faced again with a choice between Burgundy and Bordeaux, and all the causes acting upon the physical and biological level are exactly the same, we would envision exactly the same result – I travel down the path from D to Bx.

But whence the concept of repeatability? It only has meaning within an abstract time. If I conceive of the motion of the universal field as a series of instants, which is to say a series of instantaneous 3-D "Cubes," where each Cube comprises all of Space (Figure 5.4), then the duration of each instant, or the extent in time of each Cube, must become vanishingly small. What scale do I choose for the instant? Physics sees entire lifetimes of particles enduring but trillionths of a second and vastly less. I can descend scales until the duration is so minute, so infinitesimal, that my "Cube" of Space exists for so infinitesimally short a time that it has nothing left of the qualitative aspect of the perceived world. In fact, as we noted in Chapter II, I am stripping all *quality* as I descend. The buzzing fly of our normal scale of time becomes the immobile fly, becomes a cloud of electrons, becomes quark events, becomes an ensemble of strings, becomes ultimately so many algebraical relations… I end with a virtually homogeneous, featureless, quality-less Cube, then another, then another… I am close to achieving absolute repeatability, for I am absolutely near the concept of *quantity*. When I count, e.g., a set of apples, I strip each apple of any individuality, any differentiating quality; I ignore their individual differences; they are for this purpose homogeneous. I am only interested in the repeatability of my counting operation – apple 1 + apple 2 + apple 3 ….

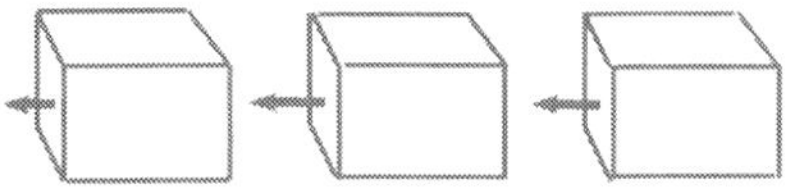

Figure 5.4. Successive Cubes of Instantaneous Space

But my (nearly) homogeneous Cubes of Space, appearing one after the other, leave me problems. The ideal limit of the extent in time of each of these Cubes is this: instantaneity. Absolute instantaneity is again my abstract, mathematical point. At this abstract, mathematical point, there is – again – no time, therefore, as we saw Lynds argue, *no change*. How then do I get Cube 1 to transition to Cube 2 which has a slightly different configuration of "particles," then to a new Cube 3, and so forth? I have achieved my repeatability. But at enormous expense. How do I get Cube 1 to generate or "cause" Cube 2? How does Cube 1 "force" the existence of Cube 2? How, in other words, do I bind the future to the present? The truth is, just as both Lynds and Bergson argued, I cannot. If the universe can be momentarily static, then it is a universe forever incapable of change.

The universe of Laplace and his demon rests upon this classical causality, which in turn rests on repeatability. I can only predict the positions and velocities of each particle at the next instant if each preceding instant is utterly homogenous and repeatable. But I can only attain absolute homogeneity by robbing my Cubes of all motion, therefore all extent in time. I end, at limit, with a universe that is incapable of change.

If the abstract notion of repeatable causes falls away, so to does the fictitious dilemma of the philosopher and his (free or not free) choice of wine. There remains the question of force itself.

Dynamical Causality, Consciousness and "Force"

There is yet the dynamic conception of the binding of present to past. It is derived directly from our conscious experience. I have an idea of an action, I initiate the action, there is throughout the act the accompanying feeling of effort or force, and the action is completed. The whole sequence, from idea to act, is experienced as a continuous whole, and by this very wholeness, the act feels prefigured in the idea. Yet it is also our deepest experience that this is not a necessary connection – the act can be aborted. It is the source of our notion that psychological causes are "different" from physical causes. But this dynamic picture of causality can in fact easily extend to the material universal field flowing in time – similar to our conscious experience – and where the future is not bound by necessity to the past.

But this indubitable experience of our consciousness, with its experience of force or *effort*, is projected upon the external world in a mixture with the concept of a Laplace-like necessary or mathematical determination. The dynamic and the static become subtly, unconsciously joined. In the union, Bergson argued, the concept of force is joined with necessity or a necessary unrolling by mathematical law – *force* now *determines* effects. We then turn this upon ourselves, viewing our own consciousness through the very layer of forms we have projected. The mechanical action of one object upon another, e.g., cue ball upon 8-ball, assumes the same form as the dynamic expression of force and resultant action experienced in consciousness, while in turn this phenomenal experience itself is painted with an algebraic prefiguring and mechanical unrolling via the force that causes it.

In the classical metaphysic, we have seen that all motion is relative. An object can move from point to point across the continuum of positions, or the continuum can move beneath the object. But we have also seen that there must be *real* motion in the matter-field – stars explode, trees grow, couch potatoes get fat. How can we distinguish real motion from that which is merely relative and which becomes rest on a change of perspective? "Force" appears as the natural answer. Force is naturally seen, in our classic framework, as that which imparts "motion" to "objects." Real motion emanates from a "force." But this is the problem: force is only a function of mass and velocity, where velocity is the *rate of change of position*. Force, or f = ma, is measured by the degree of acceleration it produces in the body (or mass, m). In turn, acceleration is only the rate of change of the rate of change of position. (Velocity is the first derivative of change of position with respect to time; acceleration is but the second derivative.) We are always dealing, in other words, only with *change of position*. In other words, these movements are still relative. The force, one with these relative movements, does not escape this relativity.[14] Force is no more absolute than the movements; it cannot serve to distinguish real or absolute motion. It is not a "cause." It is, expressed in f = ma, as physics is cautious to treat it, an invariance law. Thus, as Bergson, noted:

> It is in vain, then, that we seek to found the reality of motion on a cause which is distinct from it: analysis always brings us back to motion itself.[15]

With the same caution, psychology would say that our inner causality has no relation to the mechanical effect of one object upon another. This latter can be conceived as capable of repeating in an homogeneous space, and thus expressible by law – cue-ball repeats hitting 8-ball. But consciousness flows in *real* time, its "states" can never recur; they are not truly repeatable, each "state" being a reflection of the preceding series. Further, as already noted, it must be realized that *this is true for the motion of the entire matter-field*, the field in which our individual consciousness integrally participates. Only at a sufficient level of scale can we treat this field and its events as *practically* repeatable and subject to classical laws.

We come, then, to these implications: Mind is entirely free. The very concept of physical force is *derivative*, a projected aspect of the *psychical* evolution over time of the mind/matter-field. Our action, from intent to concrete act, participates in this psychical-physical flow. It is in making this high level conception more explicit that we will grasp the capabilities of mind and the implications for the nature of voluntary action represented in studies such as that of Cooper and Erickson.

This is the koan: who is it that acts? This is the even harder problem.

Chapter V: End Notes and References

1. The modern theory of fractals or chaos is often held to obviate this Laplacean vision, for the possibility of ever specifying the initial conditions accurately is questioned. If the value of a certain initial condition happens to be pi, i.e., the infinitely repeating decimal 3.1415926.... , the exact value can never be specified. This argument, however, has not fazed the determinists. The problem goes more deeply - to the nature of time - and this is the issue that will be examined here.
2. Gunter, P. A. Y. (1969). *Bergson and the Evolution of Physics.* University of Tennessee Press. (P. 51).
3. Bergson, H. (1889). *Time and Free Will: An Essay on the Immediate Data of Consciousness*, p. 128 .
4. Weiskrantz, L. (1997). *Consciousness Lost and Found.* New York: Oxford.
5. Birren, J. (1974). Translations in gerontology - from lab to life. *American Psychologist, 10*, 808 821.
6. Lashley, K. (1951/1970). The problem of serial order in behavior. In A. Blumenthal (Ed.), *Language and Psychology*. New York: John Wiley & Sons.
7. Bergson, H. (1912/1920). Dreams, in *Mind-Energy, Lectures and Essays,* p. 129, quoting Maury, *Le Sommeil et les Reves*).
8. Bergson, H (1912/1920). Intellectual effort, in *Mind-Energy, Lectures and Essays,* pp. 217-218)
9. Libet, B. (1985) Unconscious cerebral initiative and the role of conscious will in the initiation of action. *Behavioral and Brain Sciences, 8*, 529-566.
10. Libet, B., Freeman, A., and Sutherland, K. (1999). (Eds.), *The Volitional Brain: Towards a Neuroscience of Free Will.* London: Imprint Academic.
11. Jeannerod, M. (1994). The representing brain: Neural correlates of motor intention and imagery. *Behavioral and Brain Sciences, 17*, 187-245.
12. Robbins, S. E. (2002). Semantics, experience and time. *Cognitive Systems Research*, 301-335.
13. Cooper, L. F. & Tuthill, L. (1952). Time distortion in hypnosis and motor learning. *Journal of Psychology, 34*, 67-76.
    Cooper, L. F. & Erickson, M. (1954). *Time Distortion in Hypnosis: An Experimental and Clinical Investigation.* Baltimore: Williams & Wilkins.

14. Yes, it is precisely *acceleration* in a body that Einstein used to distinguish *real* motion in his thought experiment on riding an elevator, thus beginning his theory of gravity. One must consider that this questions the very foundations of his General Theory. In Chapter VIII, we shall go deeply into the Special Theory, and though I will be tending very little to the General Theory, there will be enough there to see these foundational difficulties. As Bergson notes, if one seeks the principle of absolute motion in force, because of the inescapable relativity of movements in the abstract continuum, you are forced back to the principle of an absolute space with absolute positions, a state for which Einstein's curved space qualifies.

15. Bergson, H. (1996/1912), p. 257.

CHAPTER VI

Meditation on a Mousetrap: Evolution and Mind

The truth is that adaptation explains the sinuosities of the movement of evolution, but not its general direction, still less the movement itself.

- Bergson, *Creative Evolution*

It is an interesting question whether the functioning and evolution of the human mentality can be accommodated within the framework of physical explanation as presently conceived...

- Noam Chomsky, *Language and Mind*

The Evolutionary Machine

Whether we are contemplating radios, robots or robins, we are viewing very complex devices. For radios or robots, we know the device was created by human minds via a not well understood process called "design," and given the difficult birth of the radio, even perhaps "creative design." For robins, the evolutionary theory of Darwin tells us things are different. The universe, acting as a giant machine, employed a form of procedure or "algorithm" to produce the robin. This procedure used random conjunctions of atoms to make chemical molecules. With more random conjunctions, it produced an elementary, living "device," perhaps a proto-cell. It then used and continues to use random mutations, in conjunction with forces or events in the external environment, to effect "natural selections" which dynamically transform devices into yet different devices, resulting in things such as robins, rabbits and a rex or two of the Tyrannosaurus type.

With this giant machine, we have removed all need to design these devices, and most significantly, any form of Mind or Intelligence designing them. This view is very much in consonance with Artificial Intelligence, which envisions machine algorithms that successfully design devices without any role required for consciousness, or conscious perception. The existence of AI and its mission is very much a hidden support of evolutionary theory. Indeed, Lloyd has proposed that the universe is a vast quantum computer wherein a few simple programs were constructed via random processes, enabling the bootstrapping of the whole complex production algorithm and machinery into existence.[1] In this view, the universe is a cosmic-scale AI device.

A warning light is flashing here. With respect to this quest to remove Mind from the universe, we have already seen that there is a form of awareness, or elementary level of mind, defined throughout the matter-field. Further, this property is required to solve both these questions: who is it that sees? And how do we see? A second warning light is the fact that a fundamental attribute of the dynamic matter-field, namely its inherent motion, underlies free action. I am focused only on the first warning light in this discussion. This light should give us pause: why the rush to remove Mind from the evolutionary process and the universal field when in fact it must stay in any case to explain conscious perception? If we cannot remove the elementary property of the universal field supporting perception, it is it truly possible to remove it for the process of design?

Design is a process of *cognition*, and cognition, with its fundamental operation of analogy, we have already seen is built integrally upon perception. Cognition equally requires consciousness, a fact that is ignored by AI and cognitive science, and the reason for which is understood by neither, precisely again due to the inadequate conception of time and its relation to mind. What I am going to show here is that one of the "pillars" supporting the entire structure of evolutionary thought described above, namely the possibility of design with artificially intelligent devices (AI) which require no consciousness, cannot stand. Technically, AI design is only a psychological support pillar or comfort. Evolutionary theory proposes to remove design altogether, intelligent or not, relying on mutations and a series of environmental causes and effects for which no detailed, actual model exists. In practice, though, expositors of evolution such as Dennett, Dawkins and numerous others make an implicit appeal to AI, while showing disturbingly little grasp of the problems. In fact, the expositors of evolution have simply blundered into AI's greatest unsolved problem, that of commonsense knowledge.

The Mousetrap and the Complexity of Devices

In recent years, some consternation arose in evolutionary theoretical circles as Michael Behe, an academic biologist, challenged the possibility of the algorithmic approach to design espoused by evolution (*Darwin's Black Box*). Though Behe dealt heavily in the biochemical realm, he placed the problem initially in the intuitive context of a mousetrap. The (standard) mousetrap consists of several parts (Figure 6.1). As a functioning whole, he argued, the trap is "irreducibly complex." For the device to work as designed, all the parts must be present and organized correctly, else it does not function.

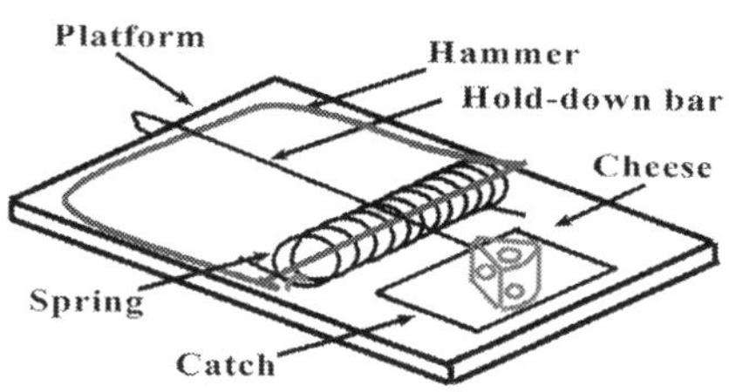

Figure 6.1. Mousetrap, M1A5, standard military issue.

The urge is to break the problem of instantiating this design into simpler components – evolving the separate, smaller parts. Natural selection buys nothing here, Behe argued. Natural selection picks some feature or form or component to continue because it happens to have been proven useful for survival. Evolving a single part (component), which by itself has no survival value, is impossible by definition – impossible, that is, by the definition of the role and function of natural selection. But even if by chance the parts evolved simultaneously, there remains the enormous problem of organization of the parts. How does this happen randomly? Each part must be oriented precisely spatially, fitted with the rest, fastened down in place, etc. There are enormous "degrees of freedom" here – ways the parts can rotate, translate and move around in space – which drive the odds against randomness to enormous proportions.

The problem can quickly be placed in the biochemical realm. Consider just one such structure in the cell alone. To manufacture palmitic acid, the cell relies on an elaborate circular molecular "machine." At the machine's center is a small arm comprised of molecules. The arm swings successively through six "workstations." Each time the arm rotates, two molecular subunits of the fatty acid are added by the action of enzymes at the workstations, and after seven rotations, the required fourteen units are present and the fatty acid released. For this rotary assembly to work, all six enzymes must be present in the right order and the molecular arm properly arranged. Now we ask, how, in what steps, always having a useful or survival value, does natural selection produce such a device?

Reviewers of Behe admit the lack of current solutions to this question. To quote one, "There are no detailed Darwinian accounts of the evolution of any fundamental biochemical or cellular system, only a variety of wishful speculations" (Shapiro, *National Review*, 1996). Nevertheless, evolutionists have reacted strongly, with attacks focusing heavily on the biological and biochemical level. An interesting case is their attack upon a favorite example used by critics of evolutionary theory involving the gas-puff firing Bombardier beetle. The beetle (there are many variants) uses a chemical combination of hydroquinones and hydrogen peroxide which collect in a reservoir. The reservoir opens into a thick-walled reaction chamber (in the beetle's rear) lined with cells that secrete catalases and peroxidases. The resulting reaction quickly brings the mixture to a boiling point, vaporizing about a fifth. The pressure closes the valve and expels the gases through openings at the tip of the abdomen in a powerful jet at a would-be attacker. If the system were not initially designed, with separate chambers for the chemicals, it is argued that the beetle himself would explode. The "exploding beetle" concept has been questioned, but more interestingly, Isaak has laid out a series of simpler beetle instantiations or steps, with examples of various steps embodied in other beetles of the class, which at least indicate a progression towards the Bombardier's sophisticated system.[2]

In sum, there are definite biological arguments for the existence of simpler stages. Note, however, that while one can demonstrate that there are simpler stages, this does not mean that one has an actual, concrete model of how one transitions from stage A to stage B, and then to stage C. More than simply the irreducible complexity of each statge, this too was the implicit force of Behe's argument. At this point, evolutionary theory invokes natural selection, which chooses B over B' or B", and which is effected by external forces of the environment. This is vague enough, while the actual creation of B, B' or B" from A requires the mechanism of mutations.

That mutations can account for change in what is called "microevolution" is unquestioned. The fish in ponds in the depths of dark caves gradually turn white. Certain light-colored moths in England during the dusty, sooty era of the industrial revolution gradually turned to a darkish color. (With the decrease of industrial pollution, they have also recently "evolved" back again to a light color.) But the assumption has been that this same mechanism can work for larger, more complex, structural transitions, where we move from dinosaur to bird, fish to frog, frog to rat, or even from variant 1 to variant 2 to variant 3 of the Bombardier beetle. This is the point of contention, and here I must discuss things at the example level of the mousetrap.

The treatment of the mousetrap example per se by evolution theorists, with its question of transitions (from device A to device B, and from B to C), is less than satisfying. In fact, as we shall see, it actually moves in the realm of AI, a realm where there are great problems precisely in this design dimension. Keep in mind that while in the biological realm, we tend to talk about these transitions simply as "mutations," there is much more going on, for just as in the mousetrap, we are talking about complex spatial fittings and fastenings of parts, complex form shaping and fabrications of the parts from materials. To effect this would require extremely complex "programming" or modifications of the sequences in the genetic instructions to bring this about – i.e., sequences of action that leave random probability behind and verge on *artificial* design.

Evolution Theorists Attack the Mousetrap

An argument, often cited as though it were a definitive critique, was provided by McDonald to demonstrate how the mousetrap could have simpler instantiations.[3] His caveat is that this is not an analogy for evolution per se, but the argument *is* taken as a critique of Behe.[4] Working backwards, McDonald gradually simplified the trap, producing four "predecessor" traps of decreasing complexity. Behe argued, however, it is not that simpler mousetraps do not exist. The question is progression – the actual mechanism of movement from A to B to C. If McDonald *is* taken as a defense of evolution, Behe easily produces a strong counter argument.[5] Starting with McDonald's first and least complex trap

(Figure 6.2, left) in the "sort of evolving" series, he examined the steps needed for McDonald to arrive at the second trap (Figure 6.2, right). The first (or single piece) trap has one arm, under tension, propped up on the other arm. When jiggled, the arm is released and comes down, pinning the mouse's paw. It is a functional trap.

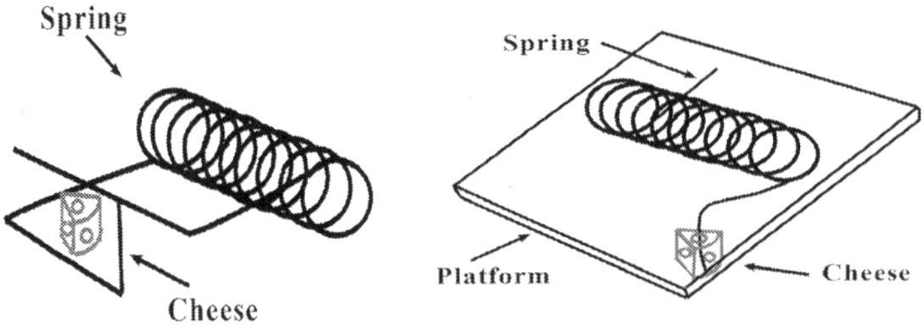

Figure 6.2. Mousetraps #1 (left) and #2 (right).

The second trap has a spring and a platform. One of the extended arms stands under tension at the very edge of the platform. If jiggled, it comes down, hopefully pinning some appendage of the mouse. To arrive at the second, functional trap, the following appears needed:

1) Bend the arm that has one bend through 90 degrees so the end is perpendicular to the axis of the spring and points toward thc platform.
2) Bend the other arm through 180 degrees so the first segment is pointing opposite to its original direction.
3) Shorten one arm so its length is less than the distance from the top of the platform to the floor.
4) Introduce the platform with staples. These have an extremely narrow tolerance in their positioning, for the spring arm must be on the precise edge of the platform, else the trap won't function.

All of this must be accomplished before the second trap will function – an intermediate but non-functional (useless) stage cannot be "selected." This complicated transition is a sequence of steps that must occur coherently. With each step required, we decrease the probability of random occurrence exponentially. As a biological example, the probability of the several amino acid mutations required for the malaria organism, *P.falciparum*, to achieve immunity to the drug chloroquine, is estimated as 1 in 10^{20} (10^{20} is a 1 with 20 zeroes). Were twice this many amino acid changes (steps) required for immunity to some new drug, the chances become 1 in 10^{40}, and for thrice the steps, 10^{60}. The probability (or improbability) is clearly a function of the number of transition "steps" identified. The "four" steps in the mousetrap transition above is an arbitrary number, and indeed could be decomposed to many further sub-steps when dealing with the level of random mutations.

Each of the subsequent transitions (2->3, 3->4, 4->5) proved subject to the same argument. McDonald then produced a more refined series of traps.[6] He argued that the point was made that a complicated device can be built up by adding or modifying one part at a time, each time improving the efficiency of the device. Yet there are still problematic transformations between many of his steps.[7] For example, in the second series, the transition between a simpler spring trap (Figure 6.3, Trap Five) and one now employing a hold-down bar (Figure 6.3, Trap Six) is a visual statement of the difficulty of the problem. Even if the simpler trap were to become a biologically based analog – a largish "mouse-catcher beetle" – sprouting six legs and a digestive system for the mice it catches, the environmental events and/or mutations which take it to the next step (as in Trap Six) would be a challenge to define.

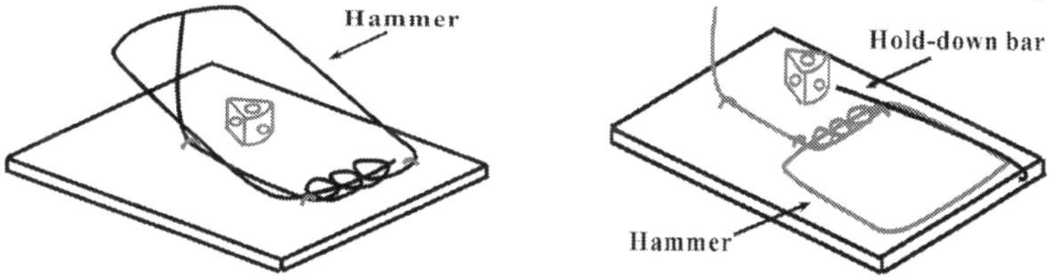

Figure 6.3. Traps Five (left) and Six (right) from the second series. Trap Six now has a hold-down bar hooked into the platform and lodged (lightly) under the hammer arm.

But the most apparently decisive evolutionary argument is that indeed biological "parts" exist that in themselves are independently functional. In essence, then, evolution has available to it pools of independently functional components from which to select, and from which to build various larger functioning wholes. Kevin Miller considered this the finding of Melendez-Hevia et al. (1996) in the realm of the Krebs cycle.[8] Miller applies this logic to the mousetrap. Each component can be conceived to be an independently functional part. For example, the hold-down bar can serve as a "toothpick," the platform as "kindling," three of the components can work together as a "tie clip" (platform, spring and hammer), and so on. The implication of this argument is disturbing, for it indicates that the grasp of the problem is deeply insufficient. Either the evolutionists, at this point, have simply become very weak AI theorists, or they know something the AI folks don't know. The fact is, this is where the evolutionists have blundered into the problem we have already met in Chapter III, that greatest of unsolved, in fact abandoned, problems of AI – *commonsense knowledge*. We shall explore this now.

The Problem of the Mousetrap in AI

Ironically, my own intellectual career, such as it is, had an early phase wherein I contemplated what it would take for an AI program to design a mousetrap.[9] This is precisely the realm of commonsense knowledge. We saw a glimpse of this in Chapter III and DORA; we go deeper and from a different angle here. The problem was presented as an initial list of components. For example, and not exhaustively, a 12" cubical box, a sharpened pencil, a razorblade, a length of string, paper clips, rubber bands, staples, toothpicks, and of course a piece of (Wisconsin) cheese. From this, the task is to create a mousetrap. (At the time, I believe this was used as a creativity test for future engineers.) One AI program I considered was Freeman and Newell's.[10] This program had a list of *functional requirements* and *functional provisions* for various objects. For example, to design a KNIFE, the program discovered that a BLADE *provided* cutting (a functional provision), but *required* holding (a functional requirement). A HANDLE *provided* holding. By matching the requirements to the provisions, the program "designed" a knife. It is precisely the implicit approach of Kevin Miller, as noted above.

I tried mightily to imagine how such a program would work in the mousetrap problem. There are many possible designs. I might make a form of crossbow, as we have seen (Figure 3.10), where the ends of the rubber band are attached to the outside of the box, the pencil (as an arrow) drawn back through a hole in the side, a paperclip holds it via a notch in the pencil, and a trip mechanism is set up with the paperclip, the string and cheese. Or I might devise a sort of "beheader," where the razorblade is embedded in the pencil as an axe (Figure 6.4), the pointed pencil-end lodged in a corner, the whole "axe" propped up by a toothpick with downward tension from the rubber band, string attached to the toothpick for a trip mechanism, etc.

What, I asked, would the database of functional provisions and requirements look like? To make the story short, I will say that I quickly abandoned any hope for this scheme. The problem is far larger. One rapidly starts to entertain the storage of "features." Noticing the "sharpness" of the pencil was integral to seeing it as supportive of the killing-function within the crossbow architecture. It is doubtful that "killing" or "piercing" would have been listed in the database as "functional provisions" of a pencil. The corner of the box provided "holding" for the pencil-axe, and while it is doubtful this would have been listed as a functional provision of box corners, it seems a type of feature. Note, meanwhile, that in the axe case, the pencil "provides" something quite different from the pencil as arrow, while a certain feature of strength and rigidity has emerged in this context.

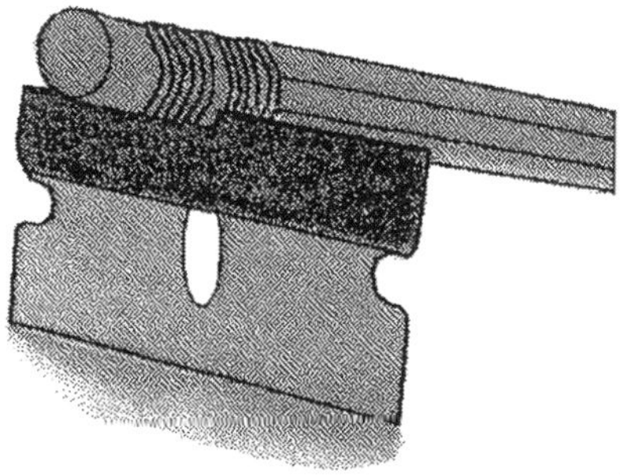

Figure 6.4. The pencil-axe.

So do we envision a list of pre-defined "features" for each object in our database? At a later date, as we have seen, in essence this would be the approach of well-known AI "analogy making" programs. But features are very ephemeral. Just like the "vertices" and "edges" of the rotating "Gibsonian" cube, they are functions of *transformations*. A fishing rod can be flexible under one transformation, sufficiently rigid under another. A floppy sock, under the appropriate "swatting" transformation, gains the rigidity and mass to become a handy fly-swatter. The pencil's rigidity under one transformation may change to just enough flexibility to support the launching of spit wads. A box may preserve its edges and corners invariant under various rotations, but lose them completely under a smashing transformation applied by the foot. And precisely the latter may be done to turn the small box in the potential components list above into a temporary dustpan. Thus we would need to store all possible transformations upon any object.

Transformations

McDonald, as we saw, performed two "bending" transformations on the wire of mousetrap #1 to obtain mousetrap #2. This form of dynamic transformation in thought heavily impressed the Gestalt psychologist, Max Wertheimer (*Productive Thinking*, 1945). He had observed children in a classroom being taught, via drawings of a parallelogram on the blackboard, the traditional, algorithmic method of dropping a perpendicular to find the area, a method which in effect turns the figure into a rectangle for easy computation of the area (length x height). Yet, when Wertheimer himself went to the board and drew a slightly rotated version of the parallelogram figure, he was shocked to see that the children failed to extend the method. They no longer knew what to do, exclaiming, "We haven't had that yet!" But outside the algorithmic-oriented classroom, Wertheimer observed a five year-old who looked at a cardboard cutout of a parallelogram, then asked for a scissors so she could cut the (triangular) end off and move it to the other side to make a rectangle (to now compute the area easily). This was bettered by the dynamic transformation exhibited by another five year-old child who formed the cardboard parallelogram into a cylinder, then asked for a scissors to cut it in half, announcing it would now make a rectangle.

It is this dynamic "folding" transformation among others that physicist and mathematician Roger Penrose (*Shadows of the Mind*) uses in examples of what he felt is "non-computational thought" – forms of thought that he felt could not be handled by the computer model of mind and its foundational concept of "computation." The formal definition of "computation" was provided by Alan Turing and embodied in an abstract computing machine he described consisting of a read-head and an infinite tape. It is now called the *Turing Machine*. All current computers, including neural nets, are simply concrete versions of a Turing Machine. This is to say that they are always doing computations that fit the definition of a Turing Machine. Obviously, if thought is truly "non-computational," it is then beyond the computer mode, for it is then beyond the computation of a Turing Machine. Thus the computer-as-mind theorists attacked Penrose mercilessly. None of his critics noticed that in his examples of non-computational thought, Penrose *had gravitated towards transformations and invariance*. That is, none of his critics grasped its significance – or wanted to.

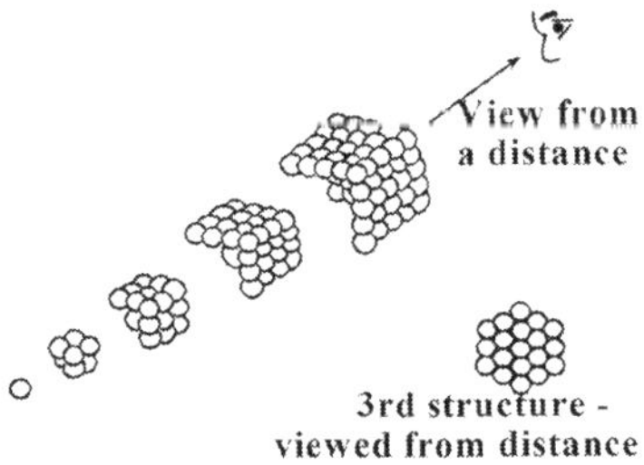

Figure 6.5. Successive cubes built from side, wall, and ceiling. Each side, wall, and ceiling structure makes a hexagonal number (1, 7, 19, 37...)

Penrose considers a proof involving "hexagonal" numbers. A hexagonal number, such as 19, can be arranged in a little hexagon such as in Figure 6.5 (bottom, right). His proof is that successive sums of hexagonal numbers are *always* a cubical number (a number that can be arranged as a cube) and hence this is "a computation that does not stop." This is to say, I can add one hexagonal number to the next and always get a cubical nmber – for an infinity of such additions. In his proof, he initially *folds* a hexagonal structure into a three-sided cube (each with a side, a wall and a ceiling). He then has us imagine building up any cube by successively stacking (another transformation) these three-faced arrangements, giving each time an ever larger cube (Figure 6.5). This is a dynamic transformation over time, in fact multiple transformations with invariants across each. We can expand the hexagonal structures successively, from 1, to 7, to 19, etc., each time preserving the visual hexagonal invariant. Then, each is folded successively, each time preserving the three-faced structural invariant. Then imagine them successively stacking, one upon the other, each operation preserving the cubical invariance. Over this event, the features (or transformational invariance) of the transformation are defined.

As a simpler example, Penrose considered how we understand that 3 x 5 = 5 x 3. Each side of the equation is different, and we can display this visually as:

3x5 (•••••) (•••••) (•••••)
5x3 (•••) (•••) (•••) (•••) (•••).

A *computational* procedure to ascertain the equality of 3x5 and 5x3 would now involve counting the elements in each group to see that we have 15 in each. But we can see this equality must be true by visualizing the array:

• • • • •
• • • • •
• • • • •

If we rotate this through a right angle in our mind's eye, we can see that nothing has changed – the new 5x3 array we see has the same number of elements as the 3x5 array pictured. We have invariance. Thus, as I noted, it is to the perception of invariance which Penrose gravitates as a natural exemplar of non-computational thought – thought which he felt requires *consciousness*. This indeed requires consciousness, we have seen, because these transformations must take place in a non-differentiable flow. These perceived invariants form his "obvious understandings" that become the building blocks for mathematical proofs. As we have seen of invariants, these obvious understandings, Penrose felt, are inexhaustible. From this he argued in effect, will arise the elements of an object language employed in a proof. But in this he was well preceded by the likes of Wertheimer (*Productive Thinking*, 1945), Arnheim (*Visual Thinking*, 1969), Bruner (*Beyond the Information Given*, 1973), Montessori (e.g., her mathematical program), Hanson (*Patterns of Discovery*, 1958), and if one looks closely, Piaget, and others.

These cases are images of *events*. The ability to represent events in the medium of an image has been utterly problematic to cognitive science. This is precisely because the origin of the mental image is simply another version of the problem of perception. If you cannot explain the origin of the *perceptual* image of the external world, you aren't going to have much luck explaining the origin of internal mental images for these images are only the retrieval of originally perceived external events, or creative combinations thereof.

Cognitive science was born with the advent of the computer metaphor of mind and brain. The essence of this model, we have seen, is that the operations of the brain can be described as the manipulations of abstract "symbols." The tone was completely set, as we saw, by Noam Chomsky in his attempt to describe the operation of human language solely in terms of symbols and rules for the manipulation of these symbols (Figure 6.6). Sentences are generated from these rules and symbols. The meaning of a sentence is supposed to be solely the end result of the set of manipulations of these symbols via the set of rules. This is the ever present abstract space and abstract time at the bottom of theoretical thinking. Time means nothing here. The manipulations are scale-less with respect to time. It makes no difference to the process how quickly or how slowly we run off our rule or set of such rules, VP = V + NP, and NP = Det + N, and S = NP + VP, etc. The end result is the same.

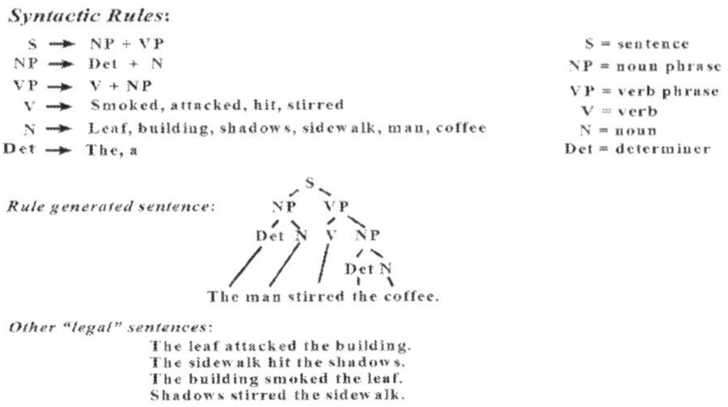

Figure 6.6 More sentences generated by syntactic rule structure. Note that these can appear semantically meaningless.

The reality is quite different. Let the sentence be, "The man stirred the coffee with the spoon." The linguistic symbol set, with its words and syntax, is simply a mediating device to move the mind to perceive an *event*. The sentence helps guide the modulation of the appropriate reconstructive wave. In this case it is the event of coffee stirring. Coffee stirring, as a dynamic event, is again defined by invariance laws and transformations (Figure 6.7). There is the radial flow field of the liquid surface being stirred. The size of the cup is constant due to the constant ratio of texture units it occludes on the table. The form of the cup is specified by its flow fields.

As we have experienced many of these coffee stirring events, other invariance laws are defined. As we discussed in the context of the origin of the "features of a cup" in Chapter III, it is as if we have recorded a "stack" of wave fronts on a hologram plate, all with

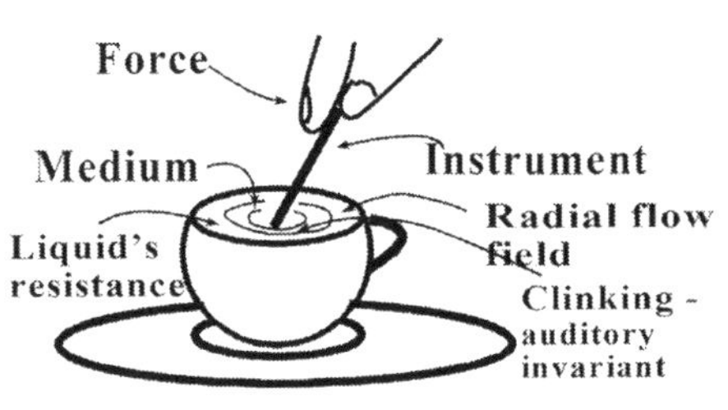

Figure 6.7. "Stirring" as an invariance structure.

nearly the same reference wave. If we pass a reconstructive wave through, the invariants are highlighted in the resulting image, the variants wash out. Some of the invariants that emerge: the cup always stays put on the table while it is stirred, the color of the cup does not change, there is a certain auditory "clink" from the spoon, there is a common circular motion of the spoon, there is an instrument with sufficient width and rigidity to move the liquid, and many, many more.

It is due to these invariance laws that we instantly recognize anomalies. Sentences like these are seen instantly as specifying an "anomalous" event:

- "As he stirred the coffee, the coffee went snap, crackle and pop,"
- "As he stirred the coffee, the cup repeatedly bulged in and out."

In effect, the phrase, "He stirred the coffee," sends a reconstructive wave through memory, redintegrating the entire "stack" of past coffee-stirring events, with their invariance laws. The rest of the event specified by the sentence simply does not "resonate" with the stack – cups don't bulge, coffee doesn't snap, crackle and pop.

Figure 6.8. A "stack" of stirring events

But in the Cognitive Science world, as well as in the robotics world, this dynamically transforming, very physical, multi-modal event, filled with forces, ratios of energy to frequency, i.e., *real dynamics*, is supposedly both stored and reconstructed as a mass of data elements (symbols) via a vast set of rules for their manipulation. One approach – the "situational calculus" – accepts the thankless task of storing all the invariance laws for events in propositional or "axiom" form. In the axioms below, we have the result specified (left side of arrow), given a certain action (stir) on the right. They say that the size, stability and elevation of the cup remain invariant while stirring.

Frame Axioms for Coffee Stirring:

Holds(s, SameSize(cup)) => Holds (Result (Stir (cup, coffee), s), SameSize(cup))

Holds(s, Stable(cup)) => Holds (Result (Stir (cup, coffee), s), Stable(cup))

Holds(s, SameElevation(cup)) => Holds (Result (Stir (cup,coffee), s), SameElevation(cup))
And so on….

This form of axiom set arose from the discovery of a fundamental problem, already noted in Chapter III, termed the *frame problem* in robotics. As the robot stirs the coffee, how does it know that the coffee cup should not change form, bulge and shrink, or change color, or the table collapse, or the sun darken, or the President of the USA get fired, or the moon crash…. The need to check the huge number of axioms in the database required to specify the appropriate state of the world is called the frame problem. The problem is compounded by the time-scale of the checks – must they happen every "instant" during the robot's action? A solution to the problem is precisely this: a way to remove the need (or significantly reduce the time) to check the frame axioms. The problem is still unresolved today. Some (and only some) AI types, I have discovered, are under the impression that it has been resolved, but this is wrong, if for no other reason than it requires a totally different form of "device."

There is a better way. Let me briefly outline it in the context of our very different "device." The anomalous event, sharing enough of the invariance structure to serve as a concrete, redintegrative cue for a stack of stirring events, can simply fail to *resonate* with a stirring event's invariance structure as defined over the retrieved stack (i.e., set of experiences). This can be *experienced* as a concretely felt, instant dissonance. Yes, the *experiencing* requires consciousness over time. Oh, we forgot, there is no role for consciousness and time in the machine framework of cognition. In any case, the former method is characteristic of the syntax-directed processor, the latter of a *semantic-directed* processor.[11] Such a semantic-directed device, however, necessarily relies on a different model of memory, namely, what we have been viewing here – an event-driven memory with its basic operation of redintegration within a non-differentiable flow of time.

In this *symbolic manipulation* framework of thought, with its frame axioms, symbols and rule manipulation schemes, etc., an influential theorist in the cognitive science field, Zenon Pylyshyn, initially denied any need for mental images, arguing that if images are simply constructed via the data elements and rules for manipulation, the image is simply redundant – all the knowledge is in the rules and data structures in the first place. What good could the image do? Images, to Pylyshyn, were redundant, and he could conceive of no need for them. Yet, strangely, they exist, for apparently no reason, at least in the computer model of mind. Thirty years later, while not denying their existence, he again challenged the field to explain why images are needed.[12] Thus, in contemplating certain "mental folding" experiments, where subjects were required to mentally fold paper into objects of certain forms, he noted that the subjects had, by necessity, to proceed sequentially through a series of folds to attain the result. Why? "Because," he argued, "*we know what happens when we make a fold.*" It has to do, he stated, with "how one's knowledge of the effects of folding is organized."

Aaron Sloman, in a seminal paper for an AI conference in 1971, already gave Pylyshyn his answer.[13] He contrasted the syntactic mode of representation (with its symbols and rules) with what he termed the *analogic* mode. In essence, this is the mode of images. In the analogic mode, there is the natural representation of *constraints* (read "laws"). Just as the cup in coffee stirring, the paper does not bulge and shrink while it is being folded. It does not disintegrate. The edges stay stable and move to overlap one another. One surface generally stays stationary. All these constraints are in fact invariance laws defined over these transformations. On the other hand, in syntactic systems, failures of reference are commonplace. The syntactically correct, "The paper screeched and burbled as it was folded," makes little semantic sense – it instantly violates an invariance across folding events. To Sloman, the greatest challenge faced by AI was achieving this (analogic) form of representation. The frame problem is simply a restatement of this problem of representational power.

We can recast Sloman's challenge: what type of "device" is required to support this form of representational power? But this is only to ask: what type of device can support perception? No visual imagery ever occurred without visual perception. The congenitally blind bear witness to this. The image is a function of, 1) perception and, 2) the memory of this perception. In turn, the image *is* the knowledge. It is no less the knowledge than the actual perceiving of an *event* of "folding" is simultaneously – knowledge. What is a "fold" other than an invariant defined over transformations in concrete experience? We have seen folds made in sheets, folds made in paper, folds made in arms/elbows, folds made in sails, folds made by Roger Penrose in three-faced hexagonal structures to make partial cubes, and even folds made with poker hands. And we have made the folds with bodily action. *Something* is always being folded. There is no such thing as an abstract "folding." "Fold" is not simply an abstract symbol among other symbols. A folding is a dynamic transformation preserving an invariant and specified in our concrete, perceptual experience.

Tunnels and Beads

AI, then, understands "Artificial;" it has an extremely impoverished grasp of the meaning of "Intelligence." Particularly, AI fails to grasp the implications of the developmental "trajectory" or path required by the child to achieve the ability to perform cognitive operations. Piaget, our oft-mentioned theorist of child cognitive development, studied and described the developmental path existing across many fairly elementary tasks. One task I wish to discuss now is the Tunnel-Bead experiment.

This simple experiment tests children aged 3-7 (*The Child's Conception of Movement and Speed*). Three beads are strung on a wire which in turn can be fitted into a small cylindrical "tunnel." The beads are of different colors, but we'll call them A, B, and C. The beads are run into the tunnel and the tunnel semi-rotated (180 degrees) from 1 to N times.

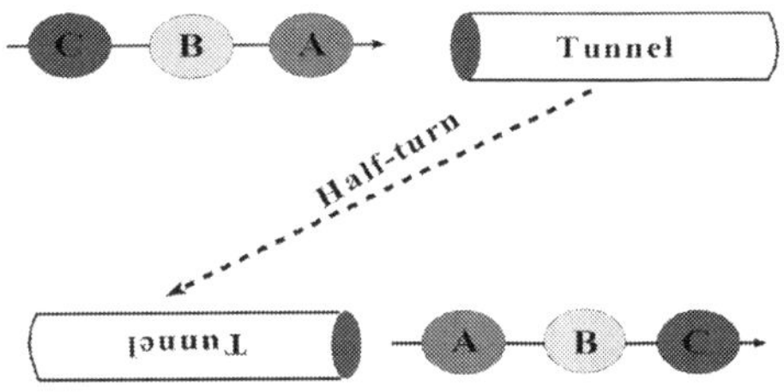

Figure 6.9. The Tunnel-Bead Experiment

A series of questions is asked, ranging from a simple, "What order will the beads come out?" after one semi-rotation (or half-turn), to the ultimate question on the order of the beads after any (n) number of half-turns. The child comes to a point of development where he can visualize the consequences of a 180^0 rotation which moves ABC to CBA and another 180^0 rotation which moves things back again to ABC, i.e., an invariance of order under a 360^0 rotation. When now asked in which order would the beads come out when the tunnel is semi-rotated 5 (or 4, or 6, or 7, etc) times, he evidences great difficulty. Some children appear to be exhausted after imagining three or possibly four semi-rotations, and they become lost when jumps are made from one number to another. As Piaget notes:

> ...But since the child, upon each half turn, endeavors to follow the inversion in every detail in his thoughts, he only gradually manages accurately to forecast the result of three, four, five half turns. Once this game of visualizing the objects in alternation is set in train, he finally discovers ...that upon each half-turn the order changes once more. Only the fact that up to this upper limit the subject continues to rely on visualizing intuitively and therefore needs to image one by one the

> half-turn, is proved because he is lost when a jump is made from one number of half-turns to any other.[14]

After this gradual perception of a higher order invariant (the "oscillation of order") defined over events of semi-rotations, there comes a point when the child can easily answer the ultimate question for the resultant order for any n-turns. Piaget's explanation, describing the "operational" character of thought, is foundational to his theory:

> Operations, one might say, are nothing other than articulated intuitions rendered adaptable and completely reversible since they are emptied of their visual content and survive as pure intention... In other words, *operations come into being in their pure state when there is sufficient schematization*. Thus, instead of demanding actual representation, each inversion will be conceived as a potential representation, like the outline for an experiment to be performed, but which is not useful to follow to the letter, even in the form of performing it mentally (emphasis added).[15]

Thus, according to Piaget, operations, freed of their imaginable content, become infinitely *compositional*. This becomes the basis for forecasting the result of n-turns, and it takes the child to about the age of seven. The operations become the generalization of actions performed through mental experiment. This is not simply abstract rules and symbols. It is not simple "rule learning." As we have seen, these "schematic" operations are built upon and do not exist without the dynamic figural transformations (images) over which invariance emerges. They are the result of a dynamical developmental trajectory required by the brain as a *self-organizing dynamic system*. This developmental path incorporates these figural transformations and requires on average *seven years*. This is the origin of the "compositional" capability of intelligence so respected, so demanded by Fodor, but so misunderstood by AI theory.

It is worth noting here that it was precisely this form of developmental trajectory, in this case over the first two years, that results in COST (Causality, Object, Space, Time) and the ability to support explicit memory. It is precisely these developmental trajectories, crucial to the self-organizing system that is the brain, that the robotics theorist seems to believe can be simply ignored, somehow manufacturing the results in one fell swoop in his robots.

Current cognitive science has persuaded itself that the computer model of mind can handle Piaget and his "abstract" operations. At this point, it should be clear that this is absurd. These operations are simply schematic transformations born of the concrete images of events – the exact thing neither AI nor computer modelers know how to account for. Beneath Piaget's operations there resides Bergson's "device."

Whether we deal with invariance over transformations, or change over transformations, we have entered the realm of time. Transformations are time-extended. Cubes are rotating. Tunnels are half-turning. Flies are buzzing by. Hexagons are "folding." As we have seen, this time-extension cannot be conceived simply as an abstract series of "instants." To support intelligence, to support cognition, we require a device that supports time-extended transformations and invariance. In other words, for cognition we require consciousness; we need a memory that spans the "instants." We need a conscious device.

But, in the previous chapters, we have seen what this "device" is.

Five Requirements for a Conscious "Device"

Let us sum up, in an abstract way, the essentials we have seen for a "device" that supports perception, and therefore, (true) cognition, and therefore the ability to design. This is the list of requirements:

1) The total dynamics of the system must be proportionally related to the events of the matter-field such that a time-scale is defined upon this field.
2) The dynamics of the system must be structurally related to the events of the matter-field, i.e., reflective of the invariance laws defined over the time-extended events of the field.
3) The operative dynamics of the system must be an integral part of the indivisible, non-differentiable motion of the matter-field in which it is embedded.
4) The information resonant over the dynamical structure (or state) must integrally include relation to or feedback from systems for the preparation of action (to ensure the partition of a subset of field events related to action).
5) The global dynamics must support a reconstructive wave.

We need a certain form of concrete wave: It is not just the organization of components in the "device," or the material from which they are made.

It is the *dynamics* they support. One does not create the concrete, electric wave of an AC generator with the proper organization of toothpicks, rubber bands or abacus beads. Whether biological or artificial, the dynamics required for perception must support a very concrete wave, establishing a ratio of proportion, i.e., a scale of time, upon the matter-field.[16]

In (3), the term "operative dynamics" is used. This is to draw the line of difference between the truly intelligent "device," and the computer model of mind. In the computer model, the effective, operative "dynamics," if you can call it that, is in the manipulation of "symbols." At the level of a computer language, this is found in the series of statements being executed, for example, "Set X = Y + 2," or "S = NP + VP." Below this higher level of language, there is another layer of symbols – simply the changing patterns of bits in memory – 1110000111 or 100100011, etc. This manipulation of symbols is simply an expression of the discrete instants and discrete objects of good old abstract space and abstract time. This Abstraction, based in an abstract space, is the basic foundation of cognitive science, and it is precisely why cognitive science and/or AI and its models have no place for consciousness in cognition, nor can either conceive of a reason why consciousness is even necessary for cognition.

I doubt that anyone knows yet how to construct a "device" that meets these five requirements. But if you want a conscious, perceiving device capable of cognition, this is what (minimally) must be done. It would be a truly intelligent robot, well beyond the conception of robots harbored by a Dennett. Unfortunately, embedded in the indivisible time-flow of the matter-field, it would be capable of free action. It might just sit in zazen for days with its koan until it sees the answer to its question, "Who is it that sees?" It would indeed be a "spiritual machine," but not the robotic machine expected by Ray Kurzweil. In fact, it might take to cracking Dennett and Kurzweil with a kyosaku as they sit in zazen, urging them to redouble their efforts towards enlightenment.

The Broadly Computational Mousetrap

We return then to the "device" underlying intelligent design. In the mousetrap task, we are designing from existing materials. I do not say from "existing components" because none of the objects are yet true components, though they have an independent function (e.g., a pencil, a rubber band). The possible "components" are being inserted into the invariance structure of an event – the drawing back and firing of a crossbow, the striking down of the axe. In the process, their requisite features emerge. This is the *analogy* defining the features.

This is a powerful transformation over a non-differentiable time. I have laid out the basis for a device with sufficient representational power to support it. As Penrose argued, it is not computational in the standard, abstract sense. This sense was laid out in the authoritative definition of *computation* given by Alan Turing. It is embodied in the definitional paradigm of all existing computers – the abstract "Turing Machine" with its infinite tape and read head. As noted, every standard computer today is a Turing Machine. The neural networks are Turing machines. This is to say that the computer is always doing a form of computation that a Turing Machine could do, where the Turing machine performs it not as efficiently, not as quickly, but just as accurately. Yet Turing specifically defined the form of computation that he would formalize in terms of *mechanical* operations. He was thinking of the ubiquitous types of computation then found everywhere – the calculations of a bank officer balancing the ledger or of a clerk computing a total cost of purchase. "Computation" consisted of the steps a human computer could carry out, a human acting mechanically *without intelligence,* i.e., *without semantics.* It was this form of computation that he would formalize in terms of the Turing machine. This captured the mechanical knowledge and calculations of the parallelogram-challenged children in Wertheimer's classroom. But Turing did not capture the form of computation of the five year-old who looked at a cardboard cutout of a parallelogram, then asked for a scissors so she could cut the (triangular) end off and move it to the other side to make a rectangle. Nor did it capture the dynamic transformation exhibited by a five year-old child who formed the cardboard parallelogram into a cylinder, then asked for a scissors to cut it in half, announcing it would now make a rectangle.

As we have viewed the form and nature of the understanding underlying that which we can term a *semantic* "computation," it is clear that the Turing concept of computation is purely derivative. By this I mean that computation, in the Turing sense, is a simply a residue, a spatialized husk of far more powerful operations of mind supporting representative thought, in turn based in the non-differentiable motion of the matter-field. The manipulation of discrete symbols in an abstract space cannot support this, nor will a dynamical device that cannot support perception. The dynamics of the device, which cognitive science has tended to view as simply supporting an abstract set of *computations*, in fact is oriented to a far more concrete purpose, as concrete as the generation of a field of force; the dynamical brain or robotic system must generate a very concrete waveform in concrete, non-differentiable time, a wave which supports a broader form of computation, broader than Turing's narrow definition, but consonant with a broader definition he did in fact leave fully open.[17] In a word, Turing computation is again a limiting case, fundamentally based in the "projection frame" of the ever underlying abstract space and abstract time in which we tend to think (and theorize), itself a derivative concept from perception and its "objects." As with physics, this frame is what must be peeled away.

Evolutionary AI

The conclusion to which this all leads is this: it is a "device" of this power, inheriting attributes of the non-differentiable time-flow of the matter-field in which it is embedded, that is required to support the design transitions posed by McDonald's mousetraps. AI, in its current form, cannot support this process. Evolution cannot rely on AI-like or machine-like algorithms for producing forms and creatures, whether mousetraps or beetles. I have noted that physicist Seth Lloyd argued that the universe is in effect, a vast, cosmic, *quantum* computer (*Programming the Universe*). Seeded with a few very simple programs by random chance, it managed to produce ever more complex programs, ultimately producing all the forms – bluebirds, bullheads and butterflies – of this world. But again, a computer, no matter how quantum, is still in the Turing class of computing machines. It is not a device of sufficient power to support even the evolution of a mousetrap.

Now, of course, evolutionary theory says that it does not rely on AI. It puts its weight on natural selection and mutations. To be clear, it must put *all* its weight on natural selection together with mutations (or "variation"). I am simply removing any temptation to go beyond this.

Unfortunately, evolution theory has already succumbed to the temptation. As an example: bacteria have a "flagellum" – a thread-like propeller that drives them though the water (Figure 6.10). This little device has a rotating axle, turning inside a bearing, driven by a molecular motor. Behe thought it another irreducibly complex device. The premier expositor of evolution, Richard Dawkins (*The God Delusion*), while ridiculing Behe to the point of impugning his motives for publishing, approvingly references Kevin Miller – the same Kevin Miller who saw no problem building mousetraps from arbitrary components. Miller identified a mechanism comprising the Type Three Secretory System (TTTS) used by parasitic bacteria for pumping toxic substances through cell walls. Since TTTS is tugging molecules through itself, it is a rudimentary version of the flagellar motor which tugs the molecules of the axle round and round. Thus, states Dawkins, evolution must have simply "commandeered" this component for the bacterial flagellum.

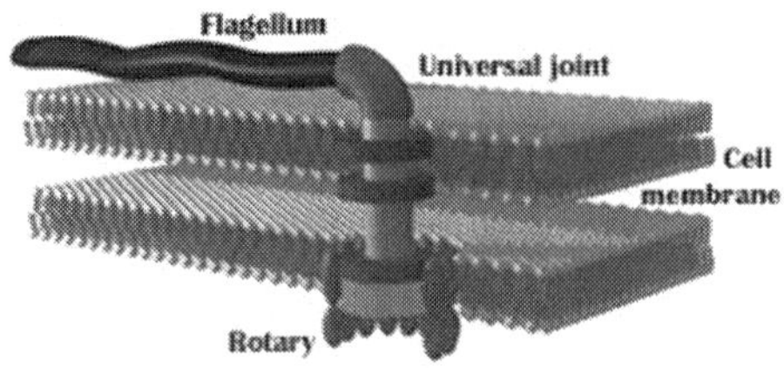

Figure 6.10. A flagellar motor.

And so the game is revealed. Just what does "commandeer" mean? Perhaps evolution's "blind watchmaker," whom Dawkins sees working by "trial and error," is peeking under his blindfold. Did evolution devise the *programs* for the selection of the components, the fittings and the transformations necessary? Then, as we have seen, evolution must be employing a far more powerful "device" than a Turing class computer. Michael Shermer (*Why Darwin Matters*) quotes Darwin's concept of "exaptation:"

> On the same principle, if a man were to make a machine for some special purpose, but were to use old wheels, springs, and pulleys, only slightly altered, the whole machine, with all its parts, might be said to be specially contrived for that purpose. Thus throughout nature almost every part of each living being has probably served, *in a slightly modified condition*, for diverse purposes, and has acted in the living machinery of many ancient and distinct specific forms. (Darwin, quoted by Shermer, my emphasis).[18]

The old "slightly modified" trick, just like slightly modifying and bending the spring-arms of the mousetrap, or adding hold-down staples and platforms, or slightly modifying the pencil to be an arrow, or to be an axe handle, or to be a pungi stake, etc., etc. Though Darwin is clearly going to be no better off than Miller in coaching AI on the design of mousetraps, in lieu of "commandeer," Shermer confidently employs the term "co-opt," as in evolution "co-opts" features to use for another purpose. For "commandeer," Eugenie Scott (*Evolution vs. Creationism*) uses "borrowing and swapping." Meanwhile, for "commandeer," Dennett (*Darwin's Dangerous Idea*) substitutes the term "generate and test," holding, with no explication, that evolution simply "generates" new devices such as flagellar motors (or mousetrap #5) to test them out. Finally, Kevin Miller himself simply uses "mix and matching" saying, "…it is to be expected that the opportunism of evolutionary processes would mix and match proteins to produce new and novel functions."[19] If Dennett, Shermer, Miller or the evolutionary biologists know secretly how to program these things, if they have solved the problem of commonsense knowledge, they should be teaching the folks in AI.[20]

Programming in Evo Devo

It may be felt that the recent discoveries of "Evo Devo" obviate these arguments.[21] It is now understood that all complex animals – people, flies, trilobites, dinosaurs and butterflies – share a common "tool kit" of master genes that govern the formation and patterning of their bodies and body parts. With this tool kit, fish fins can be modified into the legs of terrestrial vertebrates, or a simple tube-like leg can be modified into a wing. The development of these forms depends upon the turning on and off of genes at different times and places in the course of development, especially those genes that affect the number, shape and size of a structure. Further, about 3% of our DNA or roughly 100 million bits is regulatory in nature. This DNA is organized into "switches" that integrate information about position in the embryo and the time of development.

In all respects, then, we have discovered a programming language. It is a language that interfaces with the concrete, biological world, and programmed correctly, can produce complex, concrete, functioning forms. But Freeman and Newell, in their manipulation and matching of functional provisions of objects to functional requirements, also intended this to be done in a programming language. As in any complex language, its effect (its semantics) depends entirely on the correct sequencing of its instructions. It must form a proper program. If not, it either "blows up" with logic errors or produces gibberish. Unless you wish to be ridiculed by the programming profession, the complex, programmed sequence does not happen by chance, no more than the instructions of a JAVA program to display a web screen occur by luck. Some one, some thing, some force guides the sequencing derived from the complex and rich instruction set and syntax available. A flick of a "switch" to the wrong value and a leg grows on top of a fly's head – or a useless spring is placed at the wrong position on the mousetrap.

The problem posed by Behe's humble mousetrap remains in full force. Nothing has changed. The use of the language *still implies knowledge of its semantics*, and in the mousetrap context, this still involves the transformations, positioning, fittings and fastenings of parts that all work toward a concrete function. The smug rejection of mousetraps should cease, and the deep problem they represent be addressed. Until then, I expect that we still will see the liberal use of the equivalents of "co-opting" and "commandeering," now appearing in statements such as "evolution created this new instruction set," or it "modified this instruction set."

This is not to mention one other obvious fact: In making a living in the software profession, I have used many languages – JAVA, COBOL, FORTRAN, C++, Assembler, BASIC. I have yet to hear of one that was discovered just laying around, or that defined itself and published a user manual. Some one dreamt it up. Is the powerful gene/switch language an exception? If so, how did this occur?

Not Intelligent Design

So, is this an argument for Intelligent Design in evolution? It is not that simple. In *Creative Evolution*, with detailed argument, Bergson rejected both radical mechanism and finalism. In radical mechanism we see the vision, accepted by Dennett and inherent in Darwin, of the great universal machine, unrolling or unfolding its forms and creatures, with deterministic precision. An omniscient intellect, as Laplace envisioned, cognizant of the complete set of initial conditions at the birth of the universe and the deterministic laws involved, could predict everything which ultimately evolves or appears. The word "time" means nothing to this conception. It has never taken to heart the implications of the simple fact that nothing can truly repeat, where time is melodic, where each "instant" is the reflection of the whole history of change.

Finalism is Bergson's term for the conception that the universe is the result of a vast plan, an enormous idea or conception. It is simply the inverse complement of radical mechanism. Where radical mechanism drives towards the end result via its laws and initial conditions, finalism, from the other direction, draws the results irresistibly to the fulfillment of the great idea. The unforeseen creativity of real, concrete time is eclipsed. Finalism too cannot spell t-i-m-e.

Intelligent Design, if taken from the "beginning of things," is finalism. In *Creative Evolution*, Bergson explored in detail the knot of problems presented to evolutionary theory, addressed only partially today in "irreducible complexity." Irreducible complexity, as Dawkins observes, is not new, which is to say, it has been ignored for a long time. In fact, the discussion we have had on this topic does not do justice to the scope of the problems on these lines which Bergson discussed, and which formed a portion of the reasons for his rejection of Darwin. It was with deep thought that he directed his own ship, steering a direction between finalism and mechanism. He held to a vision of evolution which respects the nature of time.

You will ask, what is this vision? I don't plan to summarize it here. A detailed attack on the theory of evolution and/or a description of an alternative is not in the scope of this little book. I am satisfied if I have removed the psychological support Darwinian evolution thinks it gains from AI and cognitive neuroscience. And simultaneously, the support cognitive science and AI think they receive from evolution. I will say this: Bergson envisioned a deep, truly creative impulse driving the time-evolution of matter, producing ever unforeseen results, yes, even to itself. He called it the Élan *Vital*.

> ...so all organized beings, from the humblest to the highest, from the first origins of life to the time in which we are, and in all places as in all times, do but evidence a single impulsion, the inverse of the movement of matter, and in itself indivisible.[22]

For this he is labeled, in knee-jerk fashion, a "vitalist" and discarded on the junk heap of philosophy. Dennett, for all his books on evolution and mind, does not seem to think Bergson existed. In fact, Bergson critiqued vitalism. But this is a philosophy that has proven less than capable of grasping the thought of a Bergson.

Dawkins would see red at this suggestion, i.e., that we might consider Bergson more deeply. He sees red at any hint that science must give up before complexity, assign the cause to God, and go home. This is the implication he sees in Intelligent Design. But I am not suggesting that anything is beyond the realm of science. What I am suggesting is that there is far more to evolution than Darwin realized, that the mechanisms are far more profound, that science has yet greater discoveries to accomplish than it currently envisions, and that it is very premature to be making conclusions about the source of the existence of this world.[23]

Chapter VI: End Notes and References

1. Lloyd, S. (2006). *Programming the Universe*. New York: Alfred A. Knopf.
2. Isaak, M. (1997). Bombardier Beetles and the Argument of Design. http://www.talkorigins.org/faqs/bombardier.html
3. McDonald, J. (2000). A reducibly complex mousetrap. //http://udel.edu/~mcdonald.oldmousetrap.html
4. Miller, K. (2003). Answering the biochemical argument from design. In N. Manson (Ed.), *God and Design: The Teleological Argument and Modern Science*. London: Routledge., 292-307.
   Young, M., & Edis, T. (2004). *Why Intelligent Design Fails*. New Jersey: Rutgers University Press.
5. Behe, M. (2000a). A Mousetrap defended: Response to critics. http://www.arn.org/docs/behe.mb_mousetrapdefended.htm
6. McDonald, J. (2002). A reducibly complex mousetrap. //http://udel.edu/~mcdonald.mousetrap.html
7. Because (for example) simpler mousetraps are shown to exist, irreducible complexity is critiqued as vague. The two traps of Figure 6.3 however clarify the issue. Trap Five is simpler than trap Six. But each trap is irreducibly complex; each fails to work as designed without all its components.

   In some cases, the trap is indeed a slightly simpler version of the same design as trap Six of Figure 6.3 might be taken as a simpler version of a standard mousetrap which works without one of the standard trap's parts. But inevitably the simpler traps morph to different designs, no longer effecting quite the same function (e.g., trapping a paw vs. smashing the poor creature).
8. Melendez-Hevia, Waddell, & Cascante (1996). The puzzle of the krebs citric acid cycle: Assembling the pieces of chemically feasible reactions, and opportunism in the design of metabolic pathways during evolution. *Journal of Molecular Evolution*, 46, 508-520.

   Behe, however, notes that this is simply like describing the various chemical transitions of oil, from its initial raw state, to gasoline, while ignoring the origin and explanation of the various and complex machinery employed at each stage of the refinery process.
9. Robbins, S. E. (1976). Time and Memory: The Basis for a Semantic-Directed Processor and its Meaning for Education. Doctoral thesis, University of Minnesota.
10. Freeman, P., & Newell, A. (1971). A model for functional reasoning in design. *Second Int. conference on artificial intelligence*. London.
11. Robbins, S. E. (2002). Semantics, experience and time. *Cognitive Systems Research, 3*, 301-337. Pylyshyn, Z. (1973). What the mind's eye tells the mind's brain: a critique of mental imagery. *Psychological Bulletin, 80*, 1-22.
12. Pylyshyn, Z. (2002). Mental imagery: in search of a theory. *Behavioral and Brain Sciences,* 25, 157-237.
13. Sloman, A. (1971). Interactions between philosophy and artificial intelligence: the role of intuition and non-logical reasoning in intelligence. *Second Int. conference on artificial intelligence.* London.

14. Piaget, J. (1946). *The Child's Conception of Movement and Speed.* New York: Ballentine, p. 30.
15. Op. cit, p. 30.
16. Robbins, S. E. (2002). Semantics, experience and time. *Cognitive Systems Research,* 301-335.
17. Copeland, B. J. (2000). Narrow versus wide mechanism: including a re-examination of Turing's views on the mind-machine issue. *Journal of Philosophy*, XCVI, 1, 5-32.
18. Shermer, M. (2006). *Why Darwin Matters: The Case Against Intelligent Design.* Times Books, p. 68.
19. Miller, Kevin (2004). The flagellum unspun: The collapse of irreducible complexity. In Michael Ruse and William Dembski (eds.) *Debating Design.*

    Miller feels that since the TTTS functions in other organisms, but is a very reduced version of the arguably irreducibly complex flagellar motor, this proves that "the contention that the flagellum must be fully assembled before any of its parts can be useful is obviously incorrect." This is a bad miscontrual of irreducible complexity, not a counter example at all. Firstly, Miller had already argued that evolution can use components functional in their own right, in the mousetrap case, say, pencils or paper-clips. This does not mean that the resulting device is not irreducibly complex, requiring all such functional components to be present simultaneously in a new configuration to work. A trigger mechanism is irreducibly complex, but so is the rifle that uses it.
20. I cannot resist a comment here. As the founding editor of *Skeptic* magazine, Shermer's skepticism is most curious. Is he skeptical about global warming? Nope. Is he skeptical about artificial intelligence, 2045, and Kurzweil's Singularity? Nope. Is he skeptical about the US government's 911 story? Nope. Is he skeptical about evolution theory's explanatory power? Nope. Is he skeptical of the academic version of earth's prehistory? Nope. He seems purely the defender of established or establishment stories. Of this position, he is most unskeptical.
21. Carroll, S. (2005) *Endless Forms Most Beautiful: The New Science of Evo Devo and the Making of the Animal Kingdom.* New York: Norton.
22. Bergson, H. (1907/1911). *Creative Evolution.* New York: Holt, p. 295.
23. Robbins, S. E. (2012). Meditation on a mousetrap: On consciousness and cognition, evolution, and time. *Journal of Mind and Behavior, 33*, 69-96.

CHAPTER VII

Education: The Battle for Mind

The Chairman sees the raised hand, is surprised and disturbed by it, but acknowledges it. Then the message is delivered.

Phaedrus says, “All this is just an analogy.”

Silence. Then confusion appears on the Chairman’s face. “What?” he says. The spell of his performance is broken.

“This entire description of the chariot and horses is just an analogy.”

“What?” he says again, then loudly. “It is the *truth*! Socrates has sworn to the Gods that it is the truth!”

Phaedrus replies, “Socrates himself says it is an analogy.”

Phaedrus versus the Chairman of the Department of Philosophy, University of Chicago

- *Zen and the Art of Motorcycle Maintenance*

My First Computer Guru

In 1977, I packed up my little family, which consisted of Sibyle, my long-suffering wife, and Elizabeth, our two-month old daughter, and set sail in our Ford Pinto, praying to avoid rear-end collisions. Our goal was Pittsburgh and Chatham College. I had attained a one-year temporary position as an assistant professor there, for which I was to teach psychological test and measurement theory, psychophysics, statistical design, advanced cognition theory, advanced memory theory, and maybe some computer courses. Not a problem. I had spent long enough working on my thesis, doing post-doc work, teaching computer labs on statistics, taking computer courses, even writing programs for statistical data crunching as a third author on a couple of test and measurement papers, to gather this improbable array.

Both the statistical design course and the test and measurement theory course were math courses. I came armed for these with a definite opinion on how they should be taught. It was based on precisely the same principle that the brain uses in perception. The essence of mathematical understanding is the perception of invariance. I had come to this understanding while teaching a lab on statistics and computer programming – an occupation that Life apparently had arranged.

It came about as I spent time in the University of Minnesota computer center, typing data on computer cards and tacking the data deck onto the end of a two foot long set of cards in a long tray comprising a complex multidimensional scaling program. Occasionally, being utterly clueless about programming, I would misplace an end-of-file card somewhere in the middle of the data. This would cause the program to blow up. Rod Rosse, who had set up the program, would arrive at the computer center every day at 3pm. Rod was several years older than I. Dressed in a white shirt and tie, slouching a bit, smoking a cigarette, he would camp at a table and spend the entire night and into the early morning in the computer center. His mission was shoving in runs of statistical programs creating what are called "Monte Carlo" distributions and examining the output. This was for his PhD thesis which was in fact a mathematical statistics thesis on properties of certain forms of statistical distributions. As far as I could determine, he existed entirely on cigarettes and coffee. As I was often bringing him my problems created in ignorance, he apparently decided the cheapest course was to take me under his wing and correct my ignorance. He was planning on teaching a lab course combining the learning of computer programming with the learning of statistics. He kindly invited me, the clueless one, to be his teaching assistant. He promised to keep me one step ahead of the class. This sounded precarious, but my ignorance was not bliss.

Rod was brilliant in statistics. He had proofs of several things in mathematical statistics – say, properties of "gamma" distributions – that he never bothered to publish. He told me once that in high school, he had been, temporarily, completely inept in algebra. He could not grasp what the other students were understanding. He had finally made a discovery. The meaning, the semantics of the algebraic equations meant nothing, nothing as far as the school or the course was concerned. One only had to learn the rules of algebraic manipulation. In other words, just learn the syntax. Just act like a syntax-directed machine. The education system was completely avoiding the semantics of, the meaning of the equations. Once freed by this insight, focusing only on learning the rules, Rod had mastered his high school algebra – and gone well beyond it. But Rod was not fooled by this. He felt that it was definitely important to learn the rules, but the meaning, the semantics of the mathematics should be taught too, else you had no idea what the rules were for, or why they worked. You will quickly come to a dead end in pursuing an advanced degree. Remembering his own experience of being lost at what was going on, he realized that deeper minds *demand meaning*, and are lost when an education system is uninterested in this dimension.

We taught programming in the FORTRAN language, a language suited ideally to programming mathematical equations. The concept here was that of engaging the students' minds in the fundamental mechanics or syntax of the mathematics defining the statistical tests. To write a computer program that computes the equations, you have to understand the syntax rules of the algebra, or the order of operations. Take the equation, x = (a + b)/c. Here the order of operations says that you must first add the terms in the parenthesis, namely, a + b. The next thing is to divide the result by c. Changing this order, for example first dividing b by c, then adding a, gives the wrong answer. For example, in x = (10 + 4)/2, then x= 7, when the equation is done in the correct order. But if I divide 4 by 2 first, and then add 10, then x = 12, which is wrong. This is exactly the kind of rule that Rod had to learn to pay attention to in high school. Rod would teach the class on Tuesday, and I would take the class through the concepts and programming assignment in more detail on Thursday. Both of us concentrated on imparting the semantics behind the math as well. The course lasted the full school year, beginning with the most basic concepts and ending with the most sophisticated of statistical tests and methods.

This was precisely the same time as Shaw's seminars. Along the way, Rod also broke out another project. He wanted to create a "syntax-directed compiler." A "compiler" is a computer program that takes the statements of a computer language, such as FORTRAN, and translates them into a "machine language" the computer can understand. To do this, the compiler must "understand" the syntax rules of the language. These are exactly the same kind of rules that we saw with Chomsky's rules (Figure 5.2) for the structure of language. The compiler is always hand-written by someone for the specific language that has to be translated to machine language. There is a compiler for FORTRAN, a compiler for COBOL, a compiler for C, etc. But a syntax-directed compiler is a cut above the normal compiler. Its task is to take in a description of a language – its syntax rules – and create a compiler for the language. In other words, it is a "compiler compiler." In theory, it could take in a description of the COBOL language and create a compiler for it, or a description of C and create a compiler for it. Rod wanted to create this compiler-compiler in FORTRAN, then a highly used language, so it could be used on any computer to create compilers for any language. To do this, we had to give FORTRAN a lot more power, to include a capability termed "list processing" and what is termed "recursion," which we did. We would work on this compiler project for another year. In truth, we ultimately saw the difficulties involved as, shall we say, too Herculean. But this is where I truly learned what a syntax-directed processor is all about.

The Variance as A-Ha Experience

But meanwhile, I worked, with difficulty, at transmitting the basis of the math to the lab class. My epiphany came one day when a female grad student and friend, JoAnne, visited my little office. She was in the required one-year graduate course on statistics, to which our lab was an adjunct. She was already well into the course, in the third quarter. She was still struggling. "I don't understand the 'variance'," she said, hoping I could help. The "variance" is a statistical concept describing distributions of data. It is fundamental, absolutely basic to statistics. For a student not to grasp this concept when already so far into the subject is an indicator that something is seriously wrong. I do not believe it was JoAnne. The educational system in many respects treats students as robots. It completely fails to grasp that understanding is based in the perception of invariance. It is precisely the perception of invariance that robots, AI and the computer models of mind will never achieve. In this failure, in the education system, the battle for mind is already being lost.

I sat down with JoAnne. I drew a tiny distribution of "test scores," with each score as a little "x". The number (N) of the scores was 12. The "range" of the scores was from 1 to 9. The average of the scores was always 5, that is, the sum of the scores divided by 12. The sum of the first set of scores was 60. This average or mean is symbolized as μ in statistics, so μ = 5. Figure 7.1 shows my first drawing. We then computed the variance for this distribution.

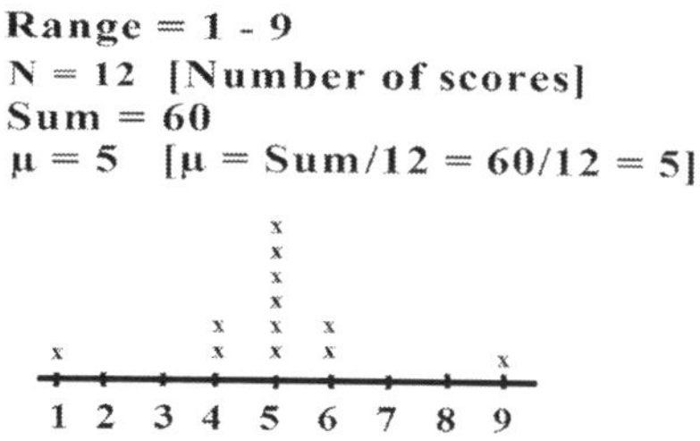

Figure 7.1. First distribution of test scores.

The variance is symbolized as σ^2 (sigma squared). The formula takes each score and subtracts it from the mean. That is, it finds the degree of difference, or distance, of each score, x, from the mean, μ. For the first score on the left, which is a 1, then 1 - 5 = -4, and for the second pair of scores, each of which is 4, 4 - 5 = -1, etc. Each difference is squared, as the squaring always makes the number positive. JoAnne and I did this for each score. The formula then adds all these differences up. The "∑" sign indicates the totaling process. The total of the differences for the distribution is 36, totaling the differences from left to right in the picture, -4^2, -1^2, -1^2, 0, 0, 0, 0, 0, 0, $+1^2$, $+1^2$, $+4^2$, all totaling 36. This total is already a rough figure for the amount of difference of the scores from the mean. The formula then divides the total of the differences by N. Figure 7.2 shows the first distribution of scores with the variance formula computed. In this case, the variance, σ^2 is 36/12 = 3.

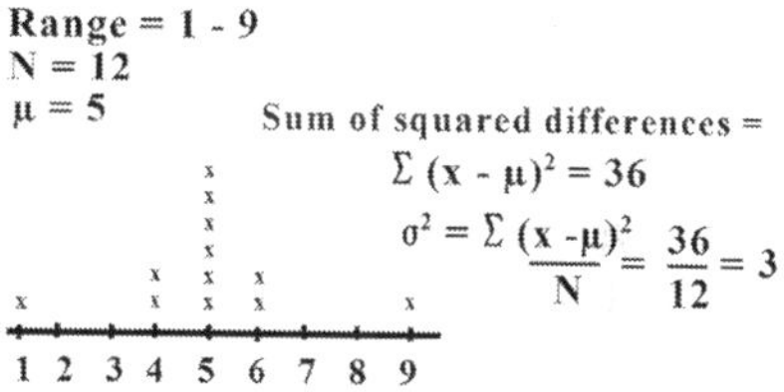

Figure 7.2. Variance, σ^2, of the first distribution. In this case, it is 3. Except for the scores at each end, this distribution of x's is clustered pretty closely around the mean score of 5.

So far, so good. But the essence of invariance is the *preservation of something over a transformation.* Displaying the transformation to the student's mind is all important. How does this formula capture a property of distributions as they transform? So JoAnne and I went though the same process for two more distributions, each time finding the differences squared, summing them up, dividing by N. Each distribution again had a range of 1-9, a mean of 5, an N of 12. These were constants across all three distributions. What differed was how the scores started to "spread out" from the mean. The figure of 7.3 shows the gradual transformation, the spreading out, of the distribution.

The scores in the distributions are gradually spreading out from the mean. Six of the twelve scores are stacked on the mean in the first distribution; only four of the twelve are stacked on the mean in the second distribution, just two of the twelve in third. For each distribution, the sum of the squared differences from the mean is growing ever larger, from 36, to 44, to 66. The variance, σ^2 then, is growing with each, from 3.0, to 3.75, to 5.16. The variance is reflecting the spreading out.

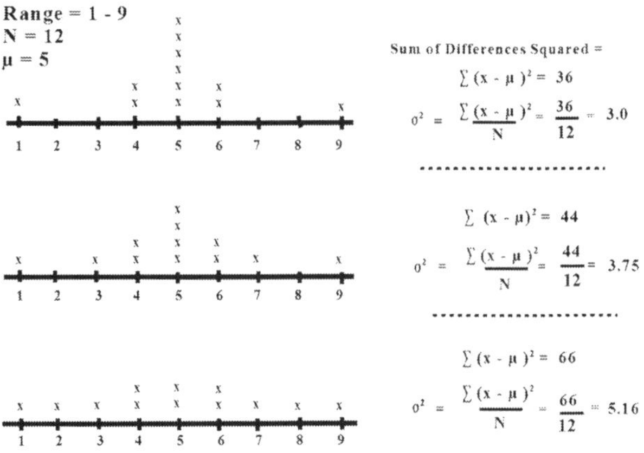

Figure 7.3. The growth of the variance, σ^2 as the scores spread out from the mean.

At the appearance of the third variance value, 5.16, for the last distribution, JoAnne jumped up and ran around the little office, exclaiming, “I understand the variance! I understand the variance!” It was her great relief and appreciation for true understanding of a hitherto meaningless mathematical fact. The education system had been focused only on imparting procedures, the rules for computing the variance, not the true *semantics* of the variance. Simply knowing the rules is a robotic “understanding,” not fit for the human mind. I applied this insight in my computer lab to ever more complex concepts in statistics. I laid this out in my doctoral thesis in its final chapter, devoted to the implications of my model of mind, for education. I argued that this is the form of true knowledge. It is based upon the perception of invariance, and it can only be supported by a “device” with the properties I have already been describing in the previous chapters of this book and which was also described in the thesis – the holographic field, reconstructive wave, etc. But this form of teaching takes work, work to work out the transformations, work to do the visuals. The best available technology then for visuals was – the blackboard. This was to be my Achilles heel at Chatham College.

Decibels vs. Dr. D

Chatham is a small woman’s college, located on beautiful grounds in Pittsburgh. My advanced perception class consisted of one girl, and the psychometric test and measurement class consisted of five girls. The latter is a very mathematically oriented class, and I began practicing my teaching philosophy. Of the five girls in the class, two were brilliant, two were struggling, and one was in between. I had not taught this particular subject before and found myself spending a lot of time at the blackboard sketching visuals. The two brilliant gals had already had an extensive course on statistics. Unfortunately it was a course taught by the purely mechanical method: "Here is the formula, here is how to compute it. This is the context to apply it. End of story." What I was aiming at is the actual intuitions, the deep understanding behind the equations – the true meaning of mathematics. The girls didn't appreciate this; what I was doing seemed to them actually unnecessary; they could do the computations. The two struggling ones, I had managed to lose. I was too green yet in getting things across.

That Christmas, the end of the fall semester, I got a pink slip. My appointment would not be extended for the next teaching year, though I had to finish out the coming semester. Chatham was big on student ratings, and apparently the girls truly did not appreciate my efforts. I would be on the market again.

The next semester, I was supposed to teach three classes – psychophysics, advanced cognition, and statistical design. Psychophysics is the study of elementary principles of perception in the various senses of hearing, sight, taste and touch, and particularly, the human sensitivity in these dimensions. It often begins with our auditory sensitivity, and this leads instantly to the decibel scale for measuring our ability to hear sounds – from very faint to very loud. The concept of the "decibel" already contains the key to psychophysical scaling. It is a ratio of an intensity of sound to a reference intensity of sound. It is logarithmic. It involves an understanding of the elementary physics of sound waves. If a student understands how this works, he has the key to psychophysics.

I had decided to change my approach. Rather than laboriously black-boarding pictures of things, I prepared a five page handout with pictures, developing the concepts – sound waves, logarithms, ratios. I gave the handout to our psychology department administrative assistant – a recent graduate of the college – to look over and to make copies. She must have been shocked. Apparently, I was "at it again." I received a call that evening from Dr. D, the head of our little three professor psychology department. I was to meet him early the next morning, before the class.

Dr. D was an older man, stooped, very tall, with a slightly droopy eye. He appeared a little like a druid. He might have been in some past life. In fact, he was a devoted student of the great historian of Druidic lore, Robert Graves. That morning, sitting in my little office, he held my handout in his hand. "This is all very well," he said, "but it is far too much. I know these students like the back of my hand. If you want teach the concept of decibels, all you need is an example like this: Suppose you have one candle. How many more candles does it take to double the intensity? Ten. How many does it take to double the intensity again? One hundred. This is all you need to say."

This example was certainly an approach, I knew. It does convey quickly the logarithmic aspect of the scale. But it hardly did justice to the concepts and mathematics of psychophysical scaling and to the concept of the decibel scale with its ratio of intensities, or to the way the logarithmic form "compressed" the vast range of sound intensity to which the ear is sensitive. Why was he so willing to short shrift these students? Why let them look at the equations in the nice psychophysics textbook they had spent sixty dollars for and feel helpless to understand where they came from? Why allow such a frustration when understanding is perfectly within their capability? Why not actually empower them? I was in hidden shock. But Dr. D was not finished. "I will teach the class this morning," he stated. "You sit in the back and watch." Something in me now drew a line. "No," I said, "I'm not doing that. You teach the class. You can teach it the whole year." He looked a bit startled, but that was the way it stayed.

I yet taught my two classes in statistical design and advanced cognition. I prepared my classes for the next day in the morning, then worked on typing my 400 page thesis into a book the rest of the morning. In the typewriter technology of that ancient age, every other word for me became a backspace, insert eraser tape, strikeover, try again. But 400 pages would eventually come out. After each typing session, I then drove downtown to work in the Pittsburgh warehouse district by the river. There I boxed up loads of sun glasses for my boss, an importer who brought them in from Italy at 80 cents a pair, sold them to stores around the area for two dollars, who in turn sold them for anywhere between fifteen and thirty-five dollars. An interesting lesson in the world of business. I yet owed three hundred dollars to a professor at Minnesota, and this was the only way I could get the extra cash. In the statistics course, I prepared all the "transformations" for statistical concepts on handouts – t-tests, f-tests, the "power" of tests, analysis of variance, orthogonal tests. The Figures 7.1, 7.2, 7.3 give a glimpse of the idea. Nowadays, it would all be done by computer on PowerPoint. The girls, seven new ones, gradually began to understand, even appreciate, my effort at transmitting the concrete intuitions and understanding of the math.

Dynamic Pi

My cognition class had one student, Jeannette, a math education major. Jeannette was the sole beneficiary of my lectures on the nature of mind. As the semester came near its close, I eventually got to my ideas on mathematics understanding as the perception of invariance. As a simple example, I told her, take the concept of pi. For many of us, it was presented in school as simply a strange mathematical fact about circles. Circles have something associated with them called pi, and pi = 3.1416. Many never penetrate beyond this. It is just another in a collection of arbitrary math facts, with little meaning beyond this or a rational origin. But the true meaning of pi is that it expresses an invariance law. As a circle expands, i.e., transforms in size, there is a constant ratio of the expanding perimeter to the expanding diameter, P/D. This constant value of P/D is pi. I told her, to teach a child pi, give her a saucer and have her measure the perimeter, and then the diameter. Then have her divide the perimeter by the diameter. Dividing the perimeter by the diameter, P/D, she will get 3.14. Now take a dinner plate and do the same. Now have her measure a wastebasket. Always, dividing the perimeter by the diameter, she will obtain 3.14. Pi is an invariant – an invariant over these transformations of circles.

The next week, Jeannette came to class and informed me, to my surprise, that she had tried my suggestion. She was student teaching a high school mathematics class. She had prepared cardboard cutouts of circles, and she had set the class to measuring. The room became a buzz of excitement as the students began to discover the invariance emerging. Pi was starting to make sense. An older teacher felt compelled to look into the room, wanting to know what was going on. Jeannette had to reassure her that things were alright. This was the good news. Jeannette was impressed and excited. To me, the bad news was that these were *seniors* in high school. This experience should have started when they five years old. As seniors, just like JoAnne, they still did not know what it meant to understand mathematics.

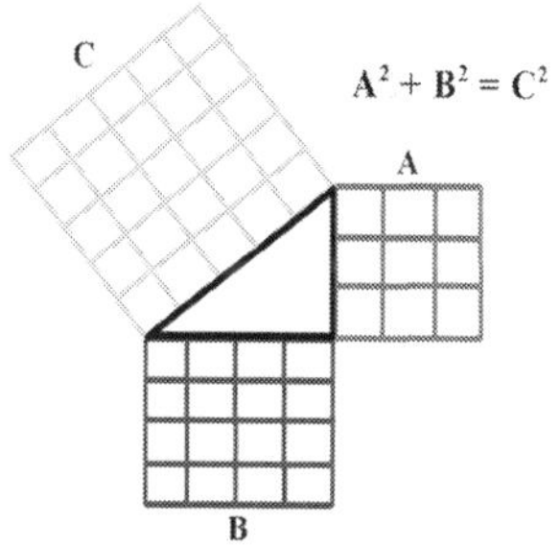

Figure 7.4. The Montessori Apparatus for the Pythagorean Theorem

It is not that certain educators have not advocated these forms of experience. Max Wertheimer, whom I have already mentioned, was one such. One can find examples occasionally in education journals. Montessori, a friend of Bergson by the way, developed a mathematical curriculum for children based entirely on the concept of providing the children concrete experience of abstract mathematical concepts. Figure 7.4 shows the Montessori apparatus, with its little block-squares, for teaching the Pythagorean Theorem, $A^2 + B^2 = C^2$. In essence, Montessori was trying to provide the basis for the concrete perception of invariance. But education has never fully accepted Montessori methods. One reason for this, strangely, is that actual, successful, empirical results mean less than theory. Montessori had no real theory of the mind backing her methods, no matter how successful. Dominating the theoretical scene since Montessori's appearance have been the useless structures of behaviorism and then the computer model of mind, neither of which can support the kind of knowledge she was intuitively striving to provide.

Wertheimer, as he observed the children in the classroom who had gained but a mechanical knowledge of the computation of the area of a parallelogram, made clear his contempt for this form of knowledge.[1] Penrose obviously sees true mathematical knowledge within the dynamic transformations and invariants of his folding cubes and other such transformations of concrete events. This is the basis for mathematical intuition, without which, a mere knowledge of facts and rules comes to a dead end in the world of higher mathematics.

Jerome Bruner (*Beyond the Information Given*, 1973) was yet another theorist that argued for this deeper form of knowledge based in concretely perceived invariance. Bruner set out to give a small group of young children around the age of eight a knowledge of quadratic equations. Children were provided with a supply of building materials. Included were some large flat squa-res of wood whose dimensions were left unspecified, the children being told that they could be considered simply x long by x wide. They also had strips of wood described as having a width of 1 and a length of x, or "1 by x". There was finally a supply of little squares, being simply "1 by 1" (Figure 7.5).The children were asked if they could make a square larger than x by x. A child might begin to make squares as in Figure 7.5.

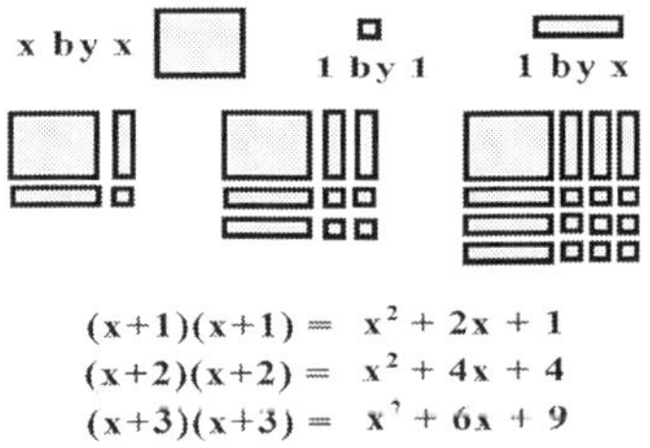

Figure 7.5. The quadratic form as a progression of squares

The child describes these squares initially very concretely – "an x square, two x strips, and a one square." He is then helped with a better symbology to write these descriptions down. Thus he writes $x^{\square}$ for an x square, 1x for the x strip, and 1 for the 1 by 1 square. The "and" is converted to a "+". So he can write $x^{\square}+2x+1$, or $x^{\square}+4x+4$. The child is asked how long and wide is a particular square, and the answer might be x+2 or an x and 2, so the whole square is $(x+2)^{\square}$. When the brackets are grasped, the child can write down his first equality: $(x+2)^{\square} = x^{\square}+4x+4$.

The child goes on making larger squares, being asked to describe each. It is close to impossible not to make structural discoveries. The child may perhaps construct a list that reveals a structure such as the following:

$x^{\square}+2x+1$ is $x+1$ by $x+1$
$x^{\square}+4x+4$ is $x+2$ by $x+2$
$x^{\square}+6x+9$ is $x+3$ by $x+3$
$x^{\square}+8x+16$ is $x+4$ by $x+4$

What is happening here, as Bruner notes, is that *syntactic* insights about regularity in notation are matched by the concrete insights gained from manipulating the physical materials.

Bruner proceeds to have the children discover the same structure using different materials, this time using a balance beam. The bar has a series of hooks numbered 1 thru 10 on each arm. Thus suppose x to be 5. Then 5 rings on hook 5 equal $x^{\square}$, 5 rings on hook 4 is 4x, and four rings on hook 1 is 4. Thus $x^{\square}+4x+4$. This he discovers can be balanced by 7 rings on hook 7, i.e., $x+2-7$, and $(x+2)^{\square} = x^{\square}+4x=4$.

Here is Bruner's appraisal of this method:

> We would suggest that learning mathematics may be viewed as a microcosm of intellectual development. It begins with instrumental activity, a kind of definition of things by doing. Such operations become represented and summarized in the form of particular images. Finally, with the help of symbolic notation that remains invariant across transformations in imagery, the learner comes to grasp the formal or abstract properties of the things he is dealing with. But while, once abstraction is achieved, the learner becomes free in a certain measure of the surface appearance of things, he nonetheless continues to rely upon the stock of imagery that permits him to work at the level of heuristic, through convenient and non-rigorous means of exploring problems and relating them to problems already encountered.[2]

But we are dealing here with more than just a "notation that remains invariant across transformations of imagery." Defined over the transformations that the child produces on the blocks are real invariants, for example the invariant defined over the progression of the square. The symbolism that the child employs is for him but the expression of this invariant in his concrete, figural experience. Thus the very meaning of these expressions is tied to invariance discovered in concrete events. Note Bruner's assertion that in essence, the ability to give meaning to mathematical expression will continue to depend, as the child progresses, on the ability to see the relation expressed as supported over imagery, i.e., the figural mode.

The fact that the quadratic form is an invariant emerging over a progression of squares points to the dynamic nature of these concepts. Arnheim, in *Visual Thinking* (1969), noted also how true understanding of a concept rests upon the perception of invariance under dynamic transformations. He pointed for example to the fact that the concept of the sum of the angles of a triangle invariantly equals 180 degrees, can only be grasped as something more than a mere fact about triangles if the particular triangle is always seen as a given instance of the transformation of two vectors (Figure 7.6). Thus the 180 degree sum of the angles is perceived as an invariant under this transformation, and the concept is inherently dynamic.

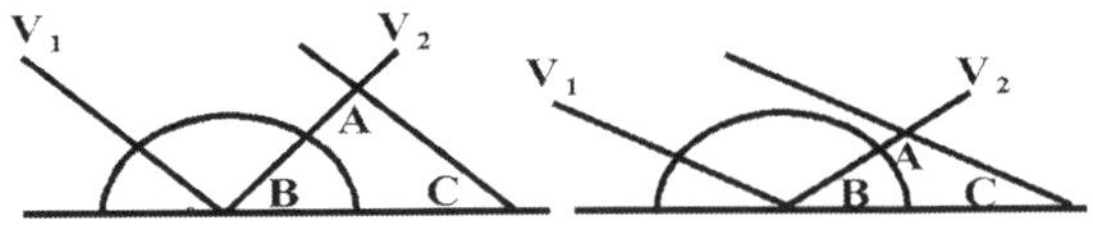

Figure 7.6. The sum of angles as 180 degrees in a triangle (ABC) as a dynamic transformation.

In my email is an article sent to me by an author, an academic PhD, for my comments. It is a proposal for a revision of K-12 mathematics education. The author noted the continuing failure of children in math, their eventual disgust with the subject, and of course, their lack of preparation to enter college. After combing through the article, I was able to determine that the author sees the core of the problem in the modern student's distrust of authority figures, while the guts of his solution for correcting this state of affairs is to teach mathematics via social media such as Twitter, because in his opinion this is the only media modern students now trust. Whatever small truth there is to this "distrust" aspect, the article reinforces not only the sad state of affairs in education, but also the sad state of affairs in theoretical analysis of the problem.

The problem started long before K-12. It started at the very beginning of the child's math education. This failure by education to grasp the true nature of mathematical understanding sits at the core of children's difficulties and their eventual hatred of mathematics. The failure is aided and abetted by the machine theory of mind which has not the wherewithal to support the role of transformations and invariance, and sits as a barrier to grasping the consistent form of teaching that is required to transmit real understanding. The achievement of true understanding is greatly self-motivating. But when children are treated as mere machines, well, the machines eventually turn themselves off.

The "Education" of the Work Place

When I left Chatham College, I entered the business world. My PhD thesis languished unpublished. I could not find a publisher willing to entertain a theory of mind so radically different from the computer model, at that point, 1978, so powerful in its ascendancy. In the world of educational psychology, I was considered too theoretical to land another college position. I found a job as a computer programmer with Mellon Bank in Pittsburgh. Mellon happened to have a mortgage division with all its software in FORTRAN, a computer language that now was an expertise of mine. I would spend thirty-three years in this profession of creating software. In the course of this time, I saw an evolution, perhaps better, a devolution of the profession and the work place in general. To my amazement, what I was witnessing was, like a slowly advancing mist across a field, the ever-growing encroachment of the machine theory of mind.

The computer systems of banks are vast affairs. The loan system, for which I was the "technical architect" for several years at my last company, I will call it Company M, consisted of several million lines of code. It held various thickets of logic – highly interconnected logic – each of which perhaps only one programmer thoroughly understood. To write this system originally took great skill. To maintain it takes great skill. To add additional functions takes great skill. Computer programming or coding is a great *skill,* just like painting pictures is a skill, or designing automobile engines, or designing mousetraps. But the essence of the onslaught of the machine theory of mind is to *de-skill.* This was summed up in a phrase that began to appear eventually in the early 90's, "coding is a commodity." By this is meant that the writing of computer programs can be reduced to an assembly line process – robots producing widgets, i.e., widgets of mindlessly produced code from specifications produced by designers. The designers of the software are somehow to have gained their design skill, miraculously, by either coming up through the robot ranks as coders mindlessly following coding specifications or else be mysteriously hatched. This philosophy was used to outsource a great proportion of the US computer programming profession to India or elsewhere. It was used to install a management structure wherein the managers are no longer organic to, or indigenous to, the profession itself, but instead are "generic" managers, managers that can manage *anything* – from developing a computer system, to creating a new drug test, to building a bridge.

The Dreyfus brothers, Hubert and Stuart, both Berkeley professors – one a professor of management, the other a philosopher of mind – saw the problem coming over twenty years ago (*Mind Over Machine*).[3] They too had attacked the computer models, with one of their perspectives being the failure of these models to capture the nature of a skill. The Swedish unions had already filed lawsuits over what they termed "de-skilling." What the Swedish already saw was the attempt to componentize a profession – break it down into tiny elements – and in the process lose the contribution of the human mind or intuition.

Mind and intuition. What have these become in the underlying thought of our current world? In the computer model of mind, there is no place for intuition. In fact there is no mind. There is certainly no grasp of the role of consciousness in thought. All models of human cognition currently proceed without any role for consciousness. The great unanswered question in cognitive science is this: what is the role of consciousness in thought? Currently, as we have seen, because real time is ignored and therefore real transformations, *consciousness has no role*.[4] This is literally to say that the models of thought or cognition "function" without any need for consciousness. The paradigm of this is the robot. Robots assemble widgets.

CMM as Group Robotics

This de-skilling process simultaneously assumes another, correlated, façade. This is that of an ever increasing tendency to *formalize* all processes. In my experience in the software world, this was an ever continuing onslaught, often led by groups in the corporation who seemed to feel that producing piles of paper is more important than a real product, that is, actual software. The quintessence of this formalistic move became the announced and heavily trumpeted goal for Company M to attain CMM level 2 and then move right on to level 3. CMM, or the Capability Maturity Model, is described, rather awkwardly, as a model of the "maturity of the capability of certain business processes." It consists, in essence, of a group of key practices divided into five levels representing the stages that an organization should go through on the way to becoming "mature." The model was born as a result of Air Force research funding in 1986 at the Carnegie-Mellon Software Engineering Institute (SEI). It was intended to create a method for evaluating software contractors and it was described in a book by Watts Humphrey, *Managing the Software Process*, in 1989. Note that Carnegie-Mellon was the home of Newell and Simon and their *Human Problem Solving*, which was in actuality about *machine* problem solving via their GPS program, and this academic institution was, and is yet, one the major forces in AI and human "cognitive simulation" via computer. Thus, it should not be too hazardous to take a guess on what is coming.

The five levels of a business that the CMM describes are:

1) Initial (chaotic, ad hoc, heroic)
2) Repeatable (project management, process discipline)
3) Defined (institutionalized)
4) Managed (quantified)
5) Optimizing (process improvement)

The CMM is touted as a general model of software process improvement. Companies are supposed to assess their current level and form a plan to ascend to the next. Level 1, for example, describes a company whose processes are supposed to be very haphazard, born at times of necessity or simply ad hoc, and relying on "heroes," where a hero is a technical professional who saves the day and the product by excessive hours of work and devotion and perhaps a programming miracle or two, or three. This level is considered eventually fatal, the sign of a doomed and/or struggling, immature company. Level 2, with its repeatable processes, is the first goal to which such a company need aspire, and which it must clamber up upon, apparently, like boarding a life raft. CMM is a widely accepted and very widely discussed and widely pushed and widely pressured model. Many companies consider it the kiss of competitive death not to have a good CMM rating. Though what I have just described appears eminently reasonable, in truth, CMM is the essence of the robotic model of mind imported into software development.

There are a few critiques of CMM, and these are remarkably referenced with great frequency in the literature. Yet their actual role and effect, given the continuing, strange CMM juggernaut, is difficult to assess. One of these, very devastating, was published in the *American Programmer* in 1994, written by James Bach, then of the highly innovative software firm of Borland.[5] Bach was obviously struck by one unbelievably contradictory fact. Not only his own company, Borland, fell in level 1, but at the time, so did Microsoft, Claris, Symantec, Oracle and Lotus, among many others. According to CMM, these and many other successful companies *should not even exist*. On the other hand, IBM, as ever a great proponent of a formalism like CMM, and themselves supposedly dwelling at an advanced "level," had managed to mangle their assignment to create a new system for the FAA's Advanced Automation project. CMM's "levels" had obviously nothing whatsoever to do with success in software development.

Bach noted that he had seen many anecdotal reports of CMM's success, but the fact is, he stated, "these could have been interpreted as evidence for the success of people working together to achieve *anything*." Eerily, exactly as I noted about the early claims of success for AI programs and even for Newell and Simon's GPS, without a comparison to some alternative model under controlled conditions, these claims are largely worthless. Ten years later, if my own experience in observing the installation of CMM in action at Company M is a typical example, such a claim would indeed be worthless. The process consisted of standardizing a large number of documents or "artifacts" of a project and making them "required." To not create a document, the Project Manager was required to do paperwork to obtain a "variance." These were documents such as Project Scopes, Project Organization charts, Requirements, Requirements Traceability, Functional Design, Technical Design, Test Plans, etc. In themselves, most of the documents could be useful, and the essential docs were in fact currently produced on a per need, context dependent basis. In many cases, they are largely useless. It is fine to standardize the form of the docs across the company. But the *religious* devotion to the production of each and every document gives the impression that CMM and its proponents believe that it is the documents that produce the project, that without them we would be lost. The project managers have been reduced to adminstrivial artists, paper pushers, gerbils running on the wheel of the CMM document production and catalogue scheme. They have few brain cycles left to consider the actual project and its problems, to think of solutions, to anticipate barriers, resource needs, testing needs, innumerable things that make a project run smoothly. In fact, to actually *lead* a project. Company M carried on despite CMM, but not without cost.

Thus, Bach noted:

> The CMM reveres process, *but ignores people*. This is readily apparent to anyone who is familiar with the work of Gerald Weinberg, for whom the problems of human interaction define engineering. By contrast, both Humphrey and CMM mention people in passing, but both also decry them as unreliable and assume that defined processes can somehow render *individual excellence* less important. The idea that process makes up for mediocrity is a pillar of CMM, *wherein humans are apparently subordinated to defined processes* (emphasis added).

We now need only to do a bit of substitution in this last, emphasized statement, taking in the context of the subject of this book. In the last phrase, for *humans*, we can substitute *minds*, and for *defined processes*, we can substitute *rule driven* (syntax-directed) *programs*. Therefore, what is being said is this: *wherein minds are subordinated to programs processing via syntactic rules*.

CMM is nothing but the AI mindset, the machine theory of mind, imported, tragically, cluelessly, into the every day world of business. All the ongoing attempts to rarify it, justify it, extend it – given this utterly false basis in a bankrupt theory of mind – are absurd.

The Attack on True Quality

Hence, this is why CMM ignores people. People are simply machines. CMM intends to turn the entire "process" into a vast machine for producing software with people as simply, unfortunately, but temporarily useful cogs and gears. So, as Bach continues:

> To render excellence less important than problem solving tasks would somehow have to be embodied in the process itself. I've never seen such a process, but if one exists, it would have to be quite complex. Imagine a process definition for playing a repeatably good chess game. Such a process exists, but it is useful only to computers; a process useful to humans has neither been documented nor taught as a series of unambiguous steps. Aren't software problems as complex as chess problems?

In other words, the "process" would have to be as complicated as a program for chess, or creating mousetraps, or creating software itself. But, as we have seen, there are no *programs* for creating mousetraps. This would require commonsense knowledge. We have seen that this knowledge is beyond the machine model of mind. Nor can the skill required to create computer software be imparted by a set of rules.

What is a skill? What is *excellence* in a skill? The Dreyfus brothers saw this as the critical issue. They outlined five stages of skill development. The five are:

1) Novice
2) Advanced Beginner
3) Competence
4) Proficiency
5) Expertise

The novice is nearest the level of machine learning. It is standard to give the novice a set of *features* to watch for, and a rule to follow based on the detection of the feature. Yes, these are the same "features" we saw feature in AI's so-called models of analogy. When learning to drive a shift car, we might be told to watch when the speedometer hits 30 mph. This is the feature to detect. It is used to determine when to apply this rule: If 30 mph, then shift from second to third. The novice chess player is given a rule which assigns point values to each piece without regard to its position on the board, and a rule that dictates "exchange your pieces for the opponent's when the total value of the those captured exceeds that of those lost." These features are termed "context free." They don't depend on the situation – whether it be the actual positions of the pieces, or the actual "strain" on the motor. This is "information processing" at its finest – context free features and rules.

The progression up the levels of skill follows a dual path. On one dimension, we learn to recognize and use ever more features, though simultaneously we are starting to use context increasingly to assess their applicability. This use of features and rules and situational modulation peaks around the "competent" level, and tapers off. Stuart Dreyfus noted that he was stuck on the "competent" level of chess skill. His academic, analytical frame of mind had made him too reliant on features and rules. He had to do too much thinking. But on the other dimension, we are learning to recognize, in a holistic fashion, an ever increasing array of patterns. As we hit the level of "proficiency," certain features of the situation stand out automatically, while others recede into the background. As the authors note:

> As events modify the salient features, plans, expectations and even the salience of features gradually changes. This happens because the proficient performer has experienced similar situations in the past and memories of them trigger plans similar to those that worked in the past and anticipations of events similar to those that occurred.[6]

They go on to note:

> A boxer seems to recognize the moment to begin an attack, not by combining by rule various elements of his body's position and that of his opponent, but when the whole visual scene in front of him and when sensations within him trigger the memory of earlier similar situations in which an attack was successful. We call the intuitive ability to use patterns without decomposing them into component features "holistic pattern recognition."[7]

What we are seeing here, in other words, is the power of redintegration, as we discussed in Chapter III – the power of invariance structures to drive or "trigger" the remembrance of events. In fact, the Dreyfus's used holograms and their recognition ability as a physical example of how this kind of instantaneous recognition could be accomplished without resort to features and rules.

At the expert level, this intuitive skill has become so much a part of him/her that he is no more aware of it than his own body. The expert driver becomes one with his car and experiences himself simply as driving rather than driving a car. The airplane pilot is no longer flying his plane, but simply experiencing flying. Chess grandmasters, engrossed in the game, lose entirely the awareness that they are manipulating chess pieces on a board, dwelling rather in a world of opportunities, hopes and fears, in which they sidestep dangers with no more effort than avoiding cars while crossing the street.Available to him/her is a vast library of board positions, estimated at 50,000, for which the desired move immediately becomes obvious. Thus, with expertise comes fluid performance. We do not think where we "place our feet," or how to "move the gas pedal." The baseball outfielder moves precisely and instantly to catch the ball. The Dreyfus's quote Taisen Deshimaru, a Japanese martial artist, "There is no choosing. It happens unconsciously, automatically, naturally. There can be no thought, because if there is thought, there is time for thought, and that means a flaw…" We have, in other words, Winograd's "situated action," or Heidegger's action without representation, his being-in-the-world.

I am not attempting to do justice here to the description of these levels of skill in *Mind Over Machine*. There is no doubt that these levels apply to the art of computer programming. Just as we discussed with the invariance structure of ecological events and the higher order invariance laws with which Vicente and Wang were dealing in baseball and chess, there is a vast world of programming structures – structures of logic, of flow, of I/O efficiency and inefficiency traps thereof, sort methods, table structuring methods, architectures for establishing general "patterns" which eliminate the need for the expensive maintenance involved in changing programs. This is not to mention the art of creating "objects" in object-oriented design that are truly useful, reusable, inheritable, etc. I have seen many expert displays of fluid skill at Company M. Greg B., a senior technician working on a "total line of credit" project with me in which we were time-pressed to bring the product to bear for a bank converting onto our loan system, simply churned out 2500 hundred lines of code in about two days – IBM estimates a programmer can produce 18 error-free lines of code a day. Greg's code created a complex algorithm for allocating a single payment across multiple loans which respected user-defined priorities and percentage schemes for the allocation, updating balances and transaction histories, and the algorithm to support the reversal of the entire process. It worked flawlessly. This, as usual, occurred with few in what had become the non-technical management structure having a clue – somewhat like children who have never touched a piano uncomprehendingly watching a virtuoso – as to the nature and skill of the performance. I could cite many, many more.

This was, of course, a *heroic* effort.[8] It was excellence. CMM hates heroes. It disparages "excellence." Remember, it holds, "defined processes can somehow render *individual excellence* less important." Why? We must go even deeper. Because excellence is Quality. The heroic is Quality. And quality, we have seen, is utterly, stupendously unaccounted for in the machine theory of mind.

One of the great expositions of this profound subject rests in Pirsig's *Zen and the Art*. In the book, Pirsig is taking a cross-county motorcycle trip with his son. Along the way, he is reconstructing the philosophical journey of Phaedrus, the name, as I noted in the introduction, that Pirsig gives to his own former personality who had suffered a psychotic episode brought on by his intense contemplations and effort after truth. He is recalling Phaedrus's explorations of Plato and Plato's war with another group of Greek philosophers whose lineage predated both Plato and Socrates – the Sophists. Plato abhors and damns the Sophists without restraint. It is not because they are in fact low and immoral people, there are far worse, but because they threaten mankind's first beginning grasp of the idea of "truth." This "truth" is in reality the foundational notion of a logic, of a reality of objective concepts that have truth independently of context or of the subjective. It is the foundation of our science, and of our computer science, and of the great Abstraction of classical space and time.

"Man is the measure of all things," said the Sophists. He is not the source of all things, as the idealists would say. Nor is he simply the passive observer of all things – there is not just a "cube" to observe. Remember Shaw's wobbly cube to which man – our brain and its constraints on the flow of the field of matter – give the "measure." "The Quality," mused Phaedrus, "which creates the world emerges as a *relationship* between man and his experience. He is a *participant* in the creation of all things – the measure of all things – it fits." But what, Phaedrus had wondered, did the Sophists mean by their term, "virtue," which they extolled above all else.

It was in reading H. D. F. Kitto's, *The Greeks*, in a passage describing "the very soul of the Homeric hero," that his realization had come. The setting is in the *Iliad*, Homer's epic poem of the siege of Troy. Troy will fall to dust; all its defenders will be killed. Hector, the Trojan prince and battle-leader, is about to go into battle. His armored form and the waving horsehair plume on his helmet have turned his infant son to crying. His wife has addressed him, voicing her fear over her fate and the loss of her husband. Hector, in replying, states his own sorrow that she may well be carried away to live in Argos, that her days of freedom will end, and she will work at the loom in another woman's house:

> And then a man will say, as he sees you weeping, 'This was the wife of Hector, who was the noblest in battle of the horse-taming Trojans, when they were fighting around Ilion.' This is what they will say, and it will be fresh grief for you, to fight against slavery bereft of a husband like that...At once shining Hector took the helmet off his head and laid it on the ground, and when he had kissed his dear son and dandled him in his arms, he prayed to Zeus and to the other gods: Zeus and ye other gods, grant that this my son may be, as I am, most glorious among the Trojans and a man of might, and greatly rule in Ilion. And may they say, as he returns from war, 'He is a far better man than his father.'

Kitto commented, "What moves the Greek warrior to deeds of heroism is not a sense of duty towards himself. He strives after that which we translate 'virtue' but is in Greek *aretê*, 'excellence'...we shall have much to say about *aretê*. It runs through Greek life." [9] When we meet *aretê* in Plato, it is translated as "virtue," carrying therefore all the ethical connotations this word now conveys. But rather, *aretê* is intended to simply mean, in all categories, *excellence*.

So Kitto continued:

> Thus the hero of the Odyssey is a great fighter, a wily schemer, a ready speaker, a man of stout heart and broad wisdom who knows he must endure without too much complaining what the gods send; and he can both build and sail a boat, drive a furrow as straight as anyone, beat a young braggart at throwing the discuss, challenge the Phaecian youth at boxing, wrestling or running; flay, skin, cut up and cook an ox, and

> be moved to tears by a song. He is in fact an excellent all-rounder; he has surpassing *aretê*.
>
> *Aretê* implies a respect for the wholeness or oneness of life, and a consequent dislike of specialization. It implies a contempt for efficiency – or rather a much higher idea of efficiency, an efficiency that exists not in one department of life, but in life itself.[10]

Phaedrus saw that Plato had appropriated the profound principle of *aretê* from the Sophists, the spiritual brethren of the great Homer, and encapsulated it, safely, as in a nicely labeled jar, as "the Good." The Good he now subordinated, in his hierarchical scheme of reality, to Truth, hence to Reason. For Plato, the Good was simply a fixed, unmoving, eternal Idea – a concept, but an utterly artificial one, not an actual concept, for we have seen that a concept, like "stirring" is actually a form of invariance over concrete experience, over the concrete flow in time of the matter-field, and this "concept" (invariance) does not exist without these concrete events, as something defined over them. For the Sophists, the Good, had they even used this label, was reality itself, ever changing ever unknowable in any fixed way – the field of matter itself in its non-differentiable, indivisible motion and therefore memory – a motion by its very nature – Quality.

At the moment Plato established this hierarchy, Quality, the essence of *aretê*, became, by definition, just another sub-category of logic. It became a minor branch of philosophy called ethics, subject to an endless, definitional game – "What is the Good? How is it defined?" In this great inversion, the machine model of mind would end up where it is today, in its vast confusion, confronted with the problem of qualia – the origin of the quality of the perceived world – and trying to explain it by the logical manipulation of symbols. Just as did the Chairman of the department of philosophy at the University of Chicago in his confrontation with Phaedrus – Phaedrus the *living* Sophist – so the machine conception of mind lies helpless before analogy, for analogy is born from this concrete flow – it is prior to all logics, it gives birth to these forms. It defines the "features" used by the machine. And CMM, also the heir of this Platonic tradition, would continue in this path, attempting to produce excellence in software by the mere installation of procedures, documents, methods and rules.

In the wings, following Plato, would come Aristotle, the quintessential representative of "the forces of formalism." Aristotle would appropriate the

Sophist subject of *rhetoric*, in essence the *concrete* process of learning – the concrete process whereby we actually learn chess by playing, learn computer programming by writing programs, learn design by having written and experienced the properties of many programs, learn auto mechanics by actually fixing cars, learn mathematics by the concrete experience of invariance. In Aristotle's hands, rhetoric became the atrocious production of endless, mind-numbing categories and distinctions:

> Rhetoric can be subdivided into particular proofs on the one hand and common proofs on the other. The particular proofs can be subdivided into methods of proof and kinds of proof. The methods of proofs are the artificial proofs and the inartificial proofs. Of the artificial proofs, there are the ethical proofs, emotional proofs and logical proofs. Of the ethical proofs, there are practical wisdom, virtue and good will. The particular methods employing artificial proofs of the ethical kind are…..

One quickly gets the idea. Phaedrus called it the "eternal smugness of the professional academician." In the classroom, it became the prototype for the mindless imparting of "facts" and "rules." Pi becomes a fact about circles. $A^2 + B^2 = C^2$ becomes a rule with no perceptual content, no grasp of the invariance law underlying it. All of algebra becomes simply a set of rules for symbol manipulation. To beings that are driven by the *semantic*, by meaning, all of this becomes inescapably – boring.

Chapter VII: End Notes and References

1. Wertheimer, M. (1945). *Productive Thinking*. New York: Harper and Row.
2. Bruner, J. (1973). *Beyond the Information Given*. New York: W. W. Norton Co., *p*. 436.
3. Dreyfus, Herbert & Dreyfus, Stuart (1986). *Mind Over Machine: The Power of Human Intuition and Expertise in the Era of the Computer*. New York: The Free Press.
4. Robbins, S. E. (2009). The COST of explicit memory. *Phenomenology and the Cognitive Sciences*, 8, 33-66.
5. Bach, James (1994). The immaturity of CMM. *American Programmer* (September).
6. Dreyfus & Dreyfus, p. 28.
7. Dreyfus & Dreyfus, p. 28.
8. A point to be made here is this: What management has failed to realize in its CMM implementing endeavors or its outsourcing endeavors is that the skill level differential of a programmer of 10-15 years experience on a large system with millions of lines of code is, relative to a novice, enormous. Just as one grandmaster can easily defeat ten novices simultaneously at chess, the output of ten novice or low level programmers can be obliterated by one experienced technician, in fact, in some cases, the ten novices would simply be *incapable* of creating the new functions or modifications correctly. The drive to create cost efficiency by outsourcing coding to a group of low-level, "inexpensive" programmers or attempting to implement a mindless CMM process using low level, robot-like programmers, is completely misguided. The non-technical management usually in charge has utterly failed to grasp the vast *leverage* and productivity represented by a set of senior, truly *skilled* technicians. This blindness, of course, is the inherent danger of the non-technical senior manager in a technical company.
9. Though we say the "Greeks" here, I would direct the reader to Iman Wilkens's brilliant *Where Troy Once Stood*. The case is indisputably made that Troy stood in England, near Cambridge. A summary is on my site: http://www.stephenerobbins.com/Other/Troy.mht
10. H. D. F. Kitto (1950). *The Greeks*. New York: Penguin.

CHAPTER VIII

The Koan of Relativity and Time

Physicists mislead us when they say there is no simultaneity. When the camera pans to the heroine tied to the rails and then to the hero rushing to the rescue on his horse – these events are simultaneous.

- James J. Gibson[1]

Bergson vs. Einstein

The theory of mind that has been described here requires a radically different model of time. This we have seen. The very nature of time is the grounding for this very mystical framework for consciousness and vision itself, where the memory intrinsic to the time-flow of the universal, holographic field allows for seeing rotating cubes and buzzing flies – and for there to be no one who is "seeing." But the theory of time is the property of physics, and in physics there is but one, dominant theory of time, and that is relativity. And this theory – nothing but an extension of the classic metaphysic – is utterly filled with misconception, misinterpretation and contradiction. We are going to expose the set of contradictions here. I refuse to let it be thought that relativity legitimately stands in the way of an enlightened theory of mind.[2]

Already, in 1922, shortly following Langevin's announcement of the twin paradox, Bergson had seen the beginnings of the misconceptions. They were worrisome enough to stir him from retirement. He wrote a book, *Duration and Simultaneity*, on the subject. The same year, 1922, Bergson engaged with Einstein in a spontaneous discussion under the auspices of the Société de Philosophie.[3] Acquiescing to an invitation to make an impromptu comment, Bergson noted, in the course of about fifteen minutes of remarks, that the concept of universal time arises from our own "proper" or experienced time in our immediate environment. He drew attention to the concept of the *simultaneity of flows*. Our experience of simultaneity, he observed, arises from our experience of multiple flows within a single flow, whether it be multiple race cars racing side by side down the track, multiple melody lines within a single flow of a symphony, multiple musicians playing on the symphony stage, multiple women cooking in the kitchen, multiple family members eating at the table, a boat floating down a river with geese flying overhead, or Gibson's hero coming to the rescue of a struggling heroine (using my own examples). This experience of multiple simultaneous flows within a single experienced flow is generalized to other perceivers, ultimately, he argued, to our concept of a *universal* flow of time. Further, this intuitive notion of simultaneity supports the very concept of relating an event to a specific time instant on a clock (as for example where an observer must relate a lightning bolt and a clock hand at 3PM as occurring simultaneously). Now, he noted, a microbe observer could say to our observer that these two events (clock hand at 3PM, lightning bolt) are not "neighboring" events at all, but are vastly distant and would not be simultaneous to a moving microbe observer. Nevertheless, to paraphrase his conclusion, he felt that this intuitive simultaneity must underlie the possibility of any time measurement at all in relativity, and was in fact the basis for reconciling the two notions.

Einstein's reply is worthy of complete quote:

> The question is therefore posed as follows: is the time of the philosopher the same as that of the physicist? The time of the philosopher is both physical and psychological at once; now, physical time can be derived from consciousness. Originally individuals have the notion of simultaneity of perception; they can hence understand each other and agree about certain things they perceive; this is a first step towards objective reality. But there are objective events independent of individuals, and from the simultaneity of perceptions one passes to that of events themselves. In fact, that simultaneity led for a long time to no contradiction [is] due to the high propagational velocity of light. The concept of simultaneity therefore passed from perceptions to objects. To deduce a temporal order in events from this is but a short step, and instinct accomplished it. But nothing in our minds permits us to conclude to the simultaneity of events, for the latter are only mental constructions, logical beings. Hence there is no philosophers' time; there is only a psychological time different from that of the physicist.[4]

This was the totality of the interchange. And so it rests. Bergson's position is, to say the least, a minority opinion. Einstein's "time of the physicist" has been the accepted criterion of reality. The simultaneity of perception is considered, at best, suspect, and in practice, invalid.

Harold Stein, a philosopher of science, essentially reprised and expanded Einstein's argument, attempting to explain ongoing misconceptions of relativity, as he saw them, in terms of our continued naïve belief in the perception of simultaneous events – an illusion based on the high velocity of light. Thus, he argued in essence, the naïve or intuitive simultaneity that perception provides is founded upon the "fleeting motions" of "masses of elements" in the brain, all subject to the limitation of communication via the velocity of light, and implying therefore that at a small enough scale of time, perceptive simultaneity would break down.[5]

This is, in fact, a curious state of affairs. Let us allow that Stein expresses Einstein's view in somewhat extended form. Then this exposition of relativity and its inherent, relativized simultaneity of events entails, or at least places a fundamental constraint upon a theory of perception. Stein is assuming a model, admittedly sketchy, of the processes in the brain underlying perception. Perception, however, is simply part and parcel of the "hard problem" of Chalmers, i.e., the explanation of conscious experience, the "world-out-there" in depth, in volume, in quality. As the problem fundamentally involves our consciousness, the problem surely cannot be divorced from our model of time. It is a problem become ever more acute, far more so than realized in Einstein's time and even just becoming so in Stein's time. Neither Stein nor Einstein could claim to have a solution. We can ask an interesting question: what if the solution to the hard problem intrinsically relies on the simultaneity of events?

We have seen that Bergson had such a solution. Sufficient it is to say that this solution contains a prediction in the sphere of perception/action that contradicts the Special Theory, though it is a contradiction if and only if physics holds that the relativization of simultaneity is a real property of time, i.e., a real, or what is termed *ontological* property of the matter-field and its temporal evolution. But this is the problem.

The "Koan" of Relativistic Effects

Let me begin with an overview of the problematic status of physical effects assigned to STR. It is a difficult topic, one which faces every student of the subject. It is in fact one of the worst of "koans," a "sound of one hand clapping" that boggles the mind, for the contradictions that arise in its interpretation are legion.

Relativity, it is well known, contains a feature which sees space units contracting and time units expanding depending on the motion of an observer. The most famous example is the twin paradox. In this case, twin Y leaves the earth at high speed in a rocket while his brother, twin X, stays on the earth. X is considered the stationary twin; he is at rest relative to Y. In motion at high velocity, Y's units of time, according to relativity, expand. Simultaneously, his space units contract. Because his time units are so much larger, he uses fewer of them, and when he returns to earth, he has aged less than his brother X. In this paradox, then, the expansion of time units and contraction of space units is considered very real. If the earth-based twin has a long beard, grey hair, and occupies a wheel chair, and the rocket-riding twin returns looking like Brad Pitt at twenty, well, we have a very real, a very physical, effect. These expansions and contractions, then, have *ontological status,* i.e., *actual being*. If this is the case, Einstein's "relativization of simultaneity" must be very real too.

What is the relativization of simultaneity? It relates to fundamental problems of measurement. Suppose, Einstein had argued, two lightning bolts strike on either side of you, fortunately a safe one thousand meters away. You happen to have two very accurate stop watches in either hand. Both are perfectly synchronized to the millisecond. You click to stop each of them when you see the light from each bolt out of the corner of your eye. You are a very fast and accurate "clicker." Behold, both watches show the same time. Further, you measure the distance from where you stood to the point where each bolt hit the ground. The distances are exactly equal. Assuming the light from each bolt traveled at the same velocity to your eyes, then the two bolts must have hit simultaneously. They traveled the same distance at the same speed, so they must have hit at the same time in order for you to have stopped both your watches at the same time. Therefore you judge these two lightening bolt events to be simultaneous. So far so good. But suppose another observer, we'll call him Observer Two, is moving on a large flying disc (his reference system) at some velocity right past where you stand (Figure 8.1). Observer Two is moving on an exact line towards the bolt on your left and away from the bolt on your right. He too has two synchronized stop watches. Note, however, that for this moving observer, the light from the bolt on the left must strike him a little sooner since he is traveling towards it, while the light from the bolt on the right gets to him a little later since he is moving away from it. He stops his two watches at different times. He declares the two-lightening bolt events *not* simultaneous.

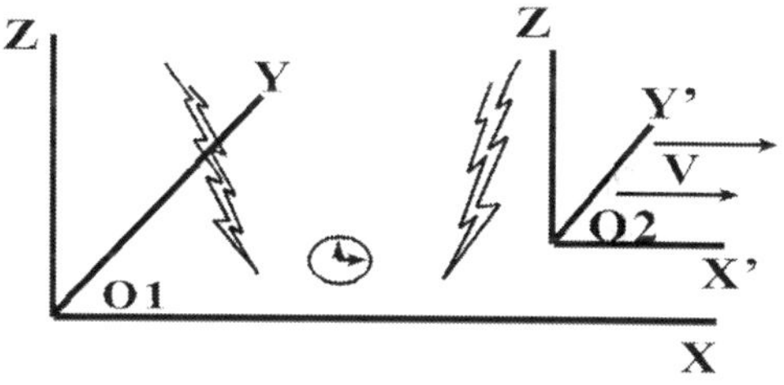

Figure 8.1. Two lightning bolts strike. Each strike at 3pm according to Observer 1 (O1) in the stationary system. Observer 2 (O2) is in the system moving at velocity, v.

Surely, we ask, he must know that he is moving! This explains the difference easily. But, said Einstein, perhaps he does not know that he is moving. Perhaps he thinks he is at rest. Perhaps he really is at rest. Perhaps it is you who are moving. How do we know? This became the essence of the first of two major postulates proposed by Einstein and which underpin his theory. The postulate is stated as, "the laws of physics are the same (invariant) in all inertial (reference) frames." It can equally be called the "reciprocity of reference systems." It implies that any observer has the right to declare himself at rest and all others in motion with respect to him. There is no way to tell who is right. The second postulate is the invariance of the velocity of light in all inertial frames.

Where do the expanded time units and contracted space units come from? Well, since Observer Two doesn't realize he is in motion (according to you); his clocks are not actually in sync. The method by which he must synchronize his clocks, Einstein showed, would be affected by his motion. One of his clocks will lag behind the other. Because of this, his measurements of distance and time within his own system will be affected. Einstein derived equations to allow us, as Observer One, to coordinate Observer Two's measurements of distances and times to our measures, in fact to specify what his measurements will look like in his system in terms of distance and time values. Central to the equations is a constant for both systems – the velocity of light. Applying these equations to Observer Two and his reference system, we would assign him expanded time units relative to ours. We would also assign him contracted distance units. At this point, one can intuitively understand why these distance and time change phenomena might be called "measurement differences." They are seeming squabbles over clock settings due to motion, but the problem of just who is in motion is very real. Observer Two, invoking reciprocity and declaring himself to be the system "at rest," can of course use the same equations for our system and for our distance and time values, claiming we are in motion and our clocks are out of sync.

Note what this implies for the simultaneity of events. The strikes of the two lightning bolts are relativized. They happen at the same time for one observer, at different times for another. Events that seem simultaneous to us may not be for another person. This means that what are simultaneous events for one observer may be successive events for another. This is to say, drilling down, that two simultaneous events for one observer, may, for another, be one event in his future, the other in his past. But what does this mean for the flow of time?

What is the classical conception of time? The advance of time traditionally involved the vision of the "time-growth" of the universe along some universally defined plane we call the "universal present." Were we to build a "space-time solid" in three-dimensions, letting the third dimension represent time, we could build one with (very thin) bread slices. Each slice represents all of 3-D space taken at an instant in time. We proceed, adding slice by slice to the "front end," gradually building a time-solid "loaf." The universal present is reduced rather mundanely to a slice of bread in this exercise. The flat surface of each slice is the universal "plane" of the present. In the classical conception, everyone's "present" is on this plane. All simultaneous events live on this plane. To us, the two lightning bolt strikes were on this plane. Any event not on this plane is either in the past, or the future – for all beings.

But now we have the relativistic fact that what are simultaneous events for one observer might be successive events for another. This implies different planes of simultaneity. It can be visualized as slices at different angles through our time-loaf. For observer X, with a plane sliced at a certain angle (Figure 8.2), certain events which he is experiencing as simultaneous events comprising his "present" can yet lie in the future for observer Y, while others lie in Y's past.

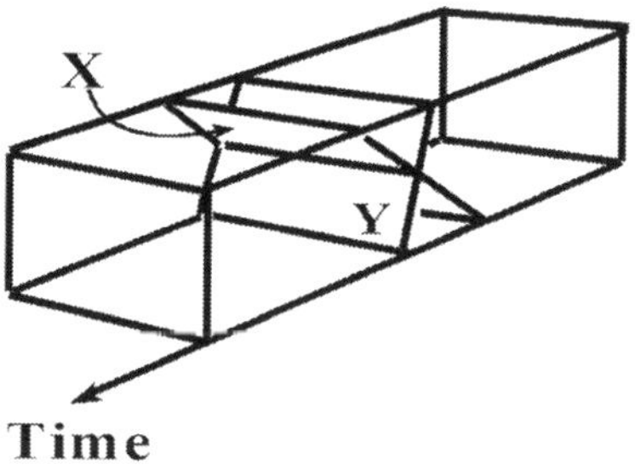

Figure 8.2. Planes of simultaneity in the space-time solid.

This vision of different futures and pasts for observers moving relative to one another makes it extremely difficult to conceive of a "universal becoming," with its vision of the growth of the universe in time along the plane of the "universal present." The conversion of simultaneities to successions, and successive events to simultaneous events, presents a troublesome difficulty for this classical conception, for the "plane of the universal present" seems to have disappeared – a single vertical slice cannot properly represent the "present."

There is, however, a natural route out of this dilemma, and it is simply to deny that there is any universal becoming, any motion of time, and to move instead to a conception of a static universe. Einstein's great collaborator, the mathematician Herman Minkowski, made statements that were the most famously conducive to this view. "Henceforth space by itself, and time by itself, are doomed to fade away into mere shadows, and only a kind of union of the two will preserve an independent reality." This conception is commonly called the "block universe." In it, there is no motion of time. All is given, past, present, future, in one giant block. This is a very common interpretation of relativistic space-time.

But let us remember, the *ontological* reality of this static block model entirely depends on the relativity of simultaneity being a fact. All depends on this relativization being a real property of the time-evolution (which we can no longer coherently visualize) of the matter-field. On this in turn depends the reality of the expanded time intervals and contracted space intervals of the rocket-riding twin Y. On this, in its turn, depends the differential aging of the twins X and Y, or the retarded aging of twin Y, as a real, physical property of matter, and the grey beards and real wrinkles.

Space Changes as Non-Ontological

When one begins to study the special theory, this is the first question that arises: are the changes in time and space real? It is extremely perplexing, for there is much to say that they are not real, and much to say they are. Here is a comment by the prolific physicist and physics writer, Paul Davies:

> How could the same thing [aging] happen at different rates?' I asked myself. I formed the impression that speed somehow distorts clock rates, so that the time dilation was some sort of illusion – an *apparent* rather than a *real* effect. I kept wanting to ask which twin experienced real time and which was deluded. ... I had to admit I could not visualize time running at two different rates and I took this to mean that I did not understand the theory. …It was then that I realized why I had been confused. So long as I could imagine the time dilation and other effects actually happening and *could work out the quantities involved, that was all that was needed.*[6]

It is not comforting to see the mechanical resolution he finally accepts, simply "doing the equations." But the contradictions are deep. Consider the initial and critical experiment to which the theory was applied, the famous 1895 experiment of Michelson and Morley. Michelson and Morley were trying to ascertain the speed of the earth through the ether. The ether was considered the all pervading, universal, fluid-like substance or medium through which energy is transmitted. Energy was considered to be propagated in waves. A wave requires some medium to ripple; in fact a wave is simply a ripple propagating through the medium. Without something like the ether, there could be no waves of energy. The earth was conceived as though it were a huge boat plowing through the ether, creating a bow wave or current. The Michelson-Morley experimental apparatus (Figure 8.3) sent out two light waves at right angles to each other. One went against the current; one went crosswise to the current.

When they ran their experiment, they obtained a strange result. The light ray running in the direction of the ether current and back should have taken longer than the light ray running crosswise. It did not; both rays took equal times. The result could be explained if the arm of the apparatus, in the direction of motion, in line with the ether flow, shrunk slightly, just enough to compensate for the theoretically larger time of travel of the light ray going though it. The light ray cheats by having a shorter course. Is such a contraction of the arm of the Michelson-Morley apparatus real, a physical fact?

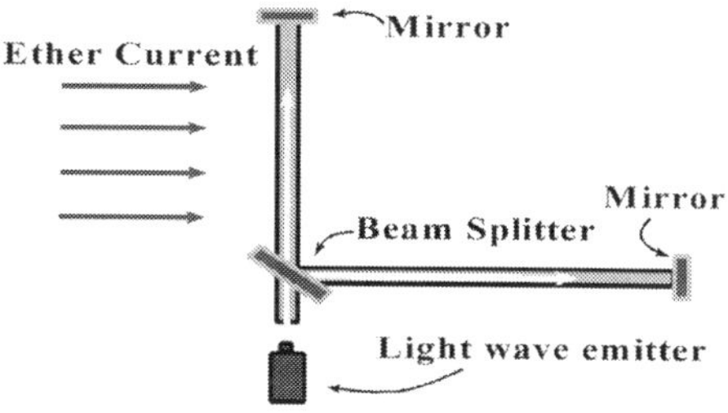

Figure 8.3. The Michelson-Morley apparatus (1895). The earth was conceived as a boat plowing thru the ether, creating an ether current or flow. The pipes/arms of the apparatus are equal in length, and an emitted light wave is split in both directions. The light wave traveling through the pipe in the direction of the current and back should have taken longer, creating an interference pattern or fringe between the two waves. However, no interference was observed; each wave takes the same time, creating a problem for the existence of the ether.

Let us remember that Hendrik Lorentz, a highly respected physicist of the time, some years before Einstein's publication, originally proposed that it was indeed real. He advanced ether-based, electro-dynamical arguments in support of equations he developed for the fore-shortening of the apparatus-arm in the direction of motion as function of velocity. His equations expressed the degree of contraction and accounted for the same travel-times. The equations looked exactly like Einstein's. But the contraction was unappealing to physics; it was rejected, or at least never accepted. Why was Einstein's "contraction," using precisely the same equations, accepted? Because the length became a space-time invariant.

How does the length become such an invariant? By being subject to the reciprocal transformations of two observers in two different reference systems, either of which can consider himself at rest and the other in motion. Einstein's perceived advance was to embed the Lorentz transformations within this symmetric, reciprocal framework, together with postulating the invariance of the velocity of light. Indeed, Einstein wished that his theory had been named "Invariantentheorie," rather than relativity.[7] In special relativity, the Lorentz transformations have no meaning with respect to just one observer. There is no invariance with just one observer. *Some form of transformation is required for an invariant.* This symmetric system is required, and within it, either observer can declare himself at rest, and then attribute the length contraction to the other (in motion), adjusting the other's space and time units to preserve the invariance of the velocity of light. Therefore as the physicist, A. P. French, states in his textbook on relativity, the length contraction is not a real property of matter, *it is a measurement effect*, "something inherent in the measurement process."[8]

In the textbooks I studied in the 1970s, the explanations of length contraction routinely told this story. The length contraction is not real. It is an effect of measurement only. The length is a space-time invariant, but no single observer has a claim on knowing the "true length." The student is warned not to fall into "the length contraction is real" trap. In truth, we must remember, there is little choice. To say that it is a real effect is to say that the Michelson-Morley apparatus arm is actually contracting somehow. This is to revert back to Lorentz and his hypothesized contraction, an explanation in fact with a real, physical model at its base – the very thing physics refused previously to accept.

Time Changes as Ontological

But as soon as the textbook turned to expanded time units or time dilation, the story was different. The problem was that there were real, physical phenomena for which time dilation appeared to be physics' only available explanation. Mesons, for example, are particles that have a certain lifespan. At rest, they exist for a certain measurable period before they decay away. When moving at high velocity, they exist for a longer period. When Lorentz's original equations are applied in this case, the increased time is perfectly predicted. Therefore time dilation is considered a quite real effect.

If there is a doubt that this is considered a very real effect, we can propose a test. We could set up a tiny electric switch a distance from the start of the meson's motion. The distance is just long enough that if the electron is not living any longer beyond its normal rest life, it won't set off the switch, but if it is living longer, it makes it to the switch and sets off an alarm clock. The ringing clock is a very real effect. Physics would quite surely accept that the meson will ring the clock.

The slow-aging Y twin with the grey and bearded X twin is simply another case of the time-dilation being considered a real effect. There is just one problem with all this. It ignores the reciprocity of reference systems. A tiny physicist on the meson should be able to say, "I'm not in motion, you are. I will never make the clock ring." The rocket-riding Y twin has perfect right to declare himself at rest, and the X twin in motion. The fact that he is on the rocket is of no account. The rocket engines could be considered to be holding the rocket's place in space as the earth moves away from the rocket, but in truth, the mathematics of relativity is abstract and these physical considerations are irrelevant. Only the abstract reciprocity of reference systems is important. So now it is the X twin who ages less. So for whom is the aging less? X or Y? Has *time* really changed? Or should we just be saying that aging period too is a space-time invariant, just as the length contraction? But fast forward. An experiment was ultimately performed in which a clock was put on a jet and flown at great speed. When the jet landed, the clock was compared to a previously synchronized counterpart left on the ground. The jet-carried clock lagged behind. The Lorentz equation for the expanded time-interval accounted for the difference – another triumph for relativity. When the experimenters stepped off the jet with their retarded clock, no one stepped forward and argued that in actuality the plane was at rest and the earth moving at extreme speed relative to the jet, thus it is the earth-based observers' clocks that should be retarded. Why not? Because obviously it is absurd. These are very real effects. They cannot be made to go away by invoking reciprocity. If the longer-living meson rings the alarm clock, the ringing is very real, it cannot be said that clock isn't ringing by suddenly remembering reciprocity. The bearded twin, should it happen, would be very real, and the beard would not go away by remembering reciprocity. The symmetry implied by reciprocity clearly has been broken.

Space Changes as Non-Ontological – Again

As far as I can ascertain, in the 1980s (perhaps earlier) another paradox began appearing in the textbooks called the "pole-barn" paradox (Figure 8.4). The "paradox" notion was now being applied to the length contraction. In this paradox, we have a longish, say, telephone pole. In its resting state, it is too long to fit into a certain barn. However, when the pole is launched into motion at a velocity near the speed of light and flies through the barn, there is a period where the pole, due its length contraction, actually fits into the barn. But *this* paradox is used as a parable for illustrating that we should *not* consider these real effects. It is unhesitatingly pointed out that the *barn* could be conceived to be in motion, and therefore the barn will contract. Now the pole does *not* fit. So the length contractions are not real, or in philosophical terms, they have no ontological status. This nicely holds the line with the interpretation of the Michelson-Morley experiment.

One could ask something however. Just like the jet-carried clock experiment, why not perform a pole-barn experiment? We could rig a mini barn-like apparatus with front-end and back-end doors that open and shut at great speed, or some analogy. The device would capture a mini-pole moving at high velocity precisely when it fits inside due to its length contraction. If we can so unhesitatingly predict that the jet-carried clock will slow down, why would we not predict that the mini-pole would contract and be trapped in the barn? But this would be admitting that the length contraction too is a very real effect. It would signal the end of any pretense of usage of the reciprocity of reference systems aspect of the special theory. At present, physics deploys the reciprocity feature for length contractions, and unhesitatingly dumps the feature for time-expansion. It therefore rejects the relativization of simultaneity as real and simultaneously (or not simultaneously?) accepts the relativization of simultaneity as real along with its block universe implication.

Those knowledgeable in this area may say, "But the twin paradox must be assigned to the General Theory (GTR)." This is due, it is thought (by some), to the accelerations involved with the rocket. Einstein's General Theory, developed after STR, deals with gravity and acceleration. This is an obviously questionable assertion on face value. If it is the twin's beard, i.e., the real, physical, obviously non-symmetric effect displayed in the aging that we are worried about, then the jet-carried clock and the meson's increased life spans must be sent to the GTR as well. These are just as real and just as non-symmetric. But I will deal with this later. Suffice it to say for now that this gambit only adds to the confusion. One quickly discovers that there is an "explanatory pea" shuffling between the General Theory and the Special Theory.

Figure 8.4. The Pole and the Barn. At high velocity, the pole fits inside the barn. At rest, it does not fit.

The Question for the Problem of Consciousness

Already a theory of consciousness has appeared (proposed by John Smythies) that assumes the standard vision of the implications of special relativity for time, namely that of the space-time block.[9] Weyl, a physicist contemporary of Einstein, expresses the implications of space-time unambiguously:

> The scene of action of reality is not a three-dimensional Euclidean space, but rather a *four-dimensional world, in which space and time are linked together indissolubly.* However deep the chasm may be that separates the intuitive nature of space from that of time in our experience, nothing of this qualitative difference enters into the objective world which physics attempts to crystallize out of direct experience. ... *Only the consciousness that passes on in one portion of this world experiences the detached piece which comes to meet it and passes behind it*, as *history.*

Weyl's statement, implying that the experienced passage of time has no objective counterpart, would have had revolutionary implications had it truly been taken to heart. But relativists themselves do not seem to have been entirely clear on the implications of the concept of space-time, and the meaning of these statements had perhaps more radical ramifications than anyone cared to make clear to anyone. We will briefly examine these.

The 'Psychical' Observer

The extensions of time-extended objects are usually called "world-lines" in relativity theory, or sometimes "tracks." "An individual," says Eddington, "is a four-dimensional object of greatly elongated form. In ordinary language, we say that he has considerable extension in time and insignificant extension in space. Practically, he is represented by a line – his track through the world."[10] The last five words – "his track through the world" – as Dunne pointed out, make his statement appear like hedging, for we must ask how the line can be both the observer and the observer's path.[11] But Eddington makes clear within the same page that the track is indeed coincident with the observer, i.e., is the observer himself. "A natural body," he says, "extends in time as well as space, and is therefore four-dimensional."

Now the first problem that presents itself is the experience of the passage of time that humanity universally shares. If everything is given, if the universe simply exists as a four-dimensional, static block of space-time, then motion has become non-existent. "Changes then correspond to individuals moving along world-lines" – this is the acknowledgment of our experience of time's motion. But just what are these individuals? To any observer viewing such a system of fixed tracks or world-lines, the appearance of motion in the dimensions representing space could be produced by the movement of a three-dimensional field of observation along a track or fourth dimension orthogonal to the other three. Thus the field would simply "come across" events (as does the 1-D field of Figure 8.5). This time-traveling field of observation we can provisionally term a "psychical" observer, for the physical observer is defined as the track traveled over. This is exactly the move Smythies accepted and utilized, envisioning "consciousness modules" moving along these tracks.

The relativists had a complex case to present, and the burden of a psychical observer, had it explicitly been acknowledged, would probably have been too much to bear. Not wanting to ignore the motion of time, however, expositors of this particular notion of space-time leave us with the non-committal statement indicating that the observer moves along his track, from which the reader may infer what he pleases. The reader usually proceeds to infer that the observer is nothing more than an organic, physical apparatus, and that this physical apparatus moves over its nebulous track in the fourth

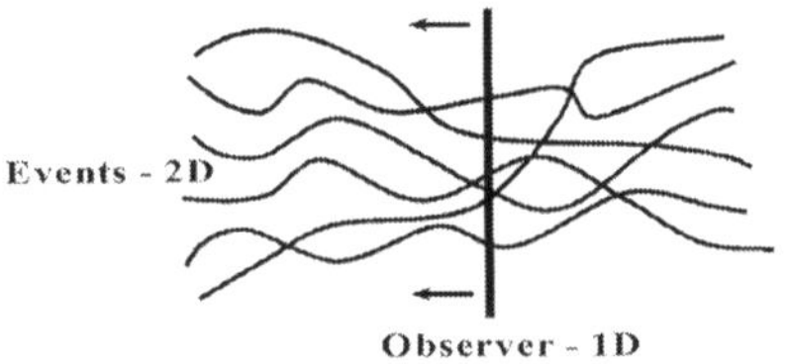

Figure 8.5. One-dimensional field traversing events in a 2-D universe

dimension. Obviously, however, a track that possessed reality to such an extent as to account for the physical characteristics of an imagined 3-D object moving along it would be, in every one of its cross-sections, physically indistinguishable from the object. Physically the track is the object extended four-dimensionally. Anything which we would consider moving along the track must differ from the track itself. Speaking of a body such as a clock or light ray moving over its track is conducive only to confusion, for the clock is physically a bundle of tracks and cannot move over itself.

Some philosophers, such as J.J.C. Smart, have noted this inconsistency.[12] Yet, respecting the static, "all is given" nature of the four-dimensional manifold, have voted solidly in favor of the concept that "there is no time." They see the passage of time as a pure illusion. Unfortunately, while they scoff at the absurdity of a psychical observer or of "consciousness running along world-lines," they offer little to put in its place. *You must at least offer a "theory of the illusion."* Even while Smart is writing his essays on time, his hand fatiguing, the ideas flowing by, he is experiencing the "illusion" in all its trickery. Whence then does the experience of the "passage" of time arise? At least the admittedly mysterious psychical observer tried to answer the question.

A Scale-less Manifold

But there is yet another thing, for we have no right to assign any particular time-scale to this manifold. We cannot envision it as it would appear to normal perception, for this perception already entails a summation over a vast history of events. If the event/world-lines the psychical observer is crossing comprise a "buzzing" fly, the choice of scales is infinite. The fly can be merely a phase in a field of vibrating strings, an ensemble of electrons/protons with no precise boundary, a fly slowly flapping his wings, or the buzzing fly of our normal perception. We would then have to account for the means whereby the time-traveling field determines scales.

Smythies would envision his traveling consciousness module as projecting a camera-like mechanism into the brain, observing the brain-tracks.[13] Again, what scale is the "camera" observing – quarks, molecular activity, chemical flows? And how are any of these – quarks or whatever –

unfolded into the world of golf balls and putting greens? This is simply what I have termed the coding problem. How is the external world of golf balls and greens unfolded from this chemical/neural/atomic code? The contents of the tracks are supposedly projected on the consciousness module's "screen." Welcome to the homunculus, observing the screen. Nor are we clear why we seem to have a whole set of observation fields moving along in parallel and constituting humanity. Why are some of us not now fighting the Peloponnesian Wars – or are we?

In any case, we could exhaust ourselves on the metaphysical, epistemological, and psychological facets of the static block reading of the implications of STR. Had psychology considered it seriously, an immediate question might have been: why are we storing memory in the brain? Clearly all events are preserved in the 4-D manifold, and the brain itself is vastly four-dimensional. If our psychic observer can go forwards, why not backwards too? Or is storage merely an illusion in the first place as we are merely coming across things that resemble past sections of the track, sections corresponding to remembering events? These and other questions might have occurred.

One might wonder how STR can pose any dilemma for a theory of consciousness when relativistic effects such as time dilation only occur at any appreciable magnitude at extremely high velocities. The normal motion velocities of organisms seem such as to make STR's effects irrelevant. However, the strange implications being noted here – the inability to account for the experienced motion of consciousness, the specter of "psychical" observers as a questionable solution to this, the curious questions about memory – are all simply functions of taking a static, four-dimensional block model of space-time seriously. This model in turn only has a possible reality if we take the relativity of simultaneity seriously (as did Smythies), i.e., as having ontological status. Proposed STR-effects such as the twin-effect, even though occurring at extremely high velocities, cement in the ontological status of these effects, and therefore the reality of the relativity of simultaneity. In the theory, the break of simultaneity begins at the most minute of velocities. Further, as we shall see when reviewing the analysis of Hagan and Hirafuji, whether or not the changes are taken as ontological, if STR is indeed valid, it places difficult constraints upon any theory of consciousness. Finally, in any case and regardless of discrepant orders of velocity, Bergson's model of perception generates a testable prediction relative to action that contradicts an implication of STR, again, only if STR's effects are taken as ontological.

Let me state this emphatically: I am not denying the reality of increased life-spans of mesons, or retarded jet-carried clocks. These phenomena are very real. The crucial question is: *how they are explained*? If changes of space and time, *as currently explained by the mathematics of relativity,* are ontological, then the relativization of simultaneity must be real. We are forced to the static block universe. A theory of consciousness is then held by this constraint, despite the difficulties into which it would inevitably place psychological theory. Given all these immensely problematic and incomprehensible implications of the static block universe for a theory of consciousness, it is time to move to a different framework of thought on the subject.

Bergson and Time

Let us remind ourselves of the heart of the difference between Bergson and Einstein. The "microbes" in Bergson's comments are an index, in essence, an index to the process of thought leading to the "objective" that Einstein must take to its logical conclusion. Bergson, in introducing them, had asked just what is the concept of "proximity" or "neighboring events" used in relativity to relate clocks to events? A microbe consciousness questions whether the clock and lightning bolt of the system of some observer are "neighboring." A micro-microbe questions the microbe's judgment of what is "neighboring;" a micro-micro-microbe does the same to the micro-microbe, and so on. Logically, we are forced to take this to its conclusion. There can be no accepted judgment of neighboring (and therefore of simultaneity) as we descend scales until we end at the mathematical point. The mathematical point is the essence of complete abstraction. The question is, is time found at all at this abstract point-event?

We saw at the foundation of Bergson's theory a critique of the *abstract* space and time implied in Einstein's vision. Abstract space, we saw Bergson arguing, is derived from the world of separate "objects" gradually identified by our perception. It ends with the notion of abstract time – the series of instants – itself simply another dimension of the abstract space. This space, argued Bergson, is in essence a "principle of infinite divisibility."

The rarifying of the abstraction continued. All motions are now treated as relative, for we can move the object across the continuum, or the continuum beneath the object. Motion now becomes immobility dependent purely on perspective. All real, concrete motion of the universal field is now lost.

But, Bergson argued, there must be *real* motion. As we saw him insist:

> Though we are free to attribute rest or motion to any material point taken by itself, it is nonetheless true that the aspect of the material universe changes, that the internal configuration of every real system varies, and that here we have no longer the choice between mobility and rest. Movement, whatever its inner nature, becomes an indisputable reality. We may not be able to say what parts of the whole are in motion, motion there is in the whole nonetheless.[14]

Bergson, we saw, would come to view the "motions" of "objects" within a global motion of this field or whole as *changes or transferences of state.* This motion is better treated in terms of a melody, the "notes" of which permeate and interpenetrate each other, the current "note" being a reflection of the previous notes of the series, all forming an organic continuity, a "succession without distinction," a motion which is indivisible. In such a global motion, there is clearly simultaneity.

The process of "objectification" which Einstein, in his response to Bergson, describes and accepts as leading us to the "real," to objective events, and which leads Stein to his "fleeting motions" of masses of "elements," is exactly the process warned of by Bergson. The "objects" of perception – purely practical partitions carved by the body's perception in the flowing universal field – are reified into the concept of abstract, independent "objects" and their "motions," and this is further rarified to "objective" space and time, with its objective, separable "events." And following this path, Einstein is consistent. These "objective," separate events are only mental constructs. They and their simultaneity are fully subject to the relativity logically inherent in their birth.

Hence, to Bergson, Einstein's "time of the physicist" is an artificial time. But this (artificial) path, we saw, is exactly the opposite of what physics has found itself to be following. The concept of abstract space and time – this "projection frame" for thought originating in perception's need for practical action – has been the obscuring layer which is slowly being peeled away.

Special Relativity and Perception

For Bergson, we saw, the perceived world is the reflection of the possibilities of bodily action. Again, succinctly, perception is *virtual action.* As noted, the fly buzzing by, his wings a-blur, is an index of the possibility of the body's action. Were the fly flapping his wings slowly, like a heron, this would be an index of a yet different possibility, in this case, reaching out slowly and grasping the fly by the wing tip. Note that in each case, this index is simultaneously reflective of a *scale of time*, also a feature of our perception.

The change of scale and form for the fly is not merely "subjective," or a "subjective modification" of experience. This is an objective effect. Virtual action, straightforwardly, makes a prediction on action relative to the increase or decrease of the velocity of underlying processes. In principle, this is a testable consequence albeit difficult today. The question is, does Special Relativity also make a prediction, and if so, what?

Let us consider the case of two observers, X and Y. We take the X system to be stationary, and Y moving relative to X at high uniform velocity. Assume there is a fly in X's system. X, at his normal velocity of processes, i.e., at his time-scale, perceives the fly as a blur. The fly, which X is observing, travels one of X's distance units using sixty wing-beats. It does this in one of X's time units, say a second. Y, moving at great velocity, has much expanded time units (and contracted space units), the time units increasing as he moves nearer to the speed of light. However, this is as X computes these units relative to his stationary system. The complementary case is Y's (in motion) view of the space-time of X. The Minkowski diagram (Figure 8.6) shows this situation. The rhombus OFGH is gradually collapsing like a scissors as the velocity of Y increases. The tangent to the hyperbola, GF, drops lower and lower below X's time unit, displaying that the time units of X, as Y sees them, are contracting steadily. Eddington had us imagine that at O, X lights up a cigar that lies along x_1 and has a very longish length of one space unit. The cigar burns one of X's units of time, being represented by the line t_1 and extending to its first unit. Y would now see the cigar as burning longer for X, in fact, as the tangent drops as v increases, it would last many units of X as assigned by Y. This could equally be X himself, aging (a form of "burning") many more time units than Y. Simultaneously, the space units of X, as Y sees them, are increasing. Thus note that GH would fall outside the space unit of X – the cigar is longer.

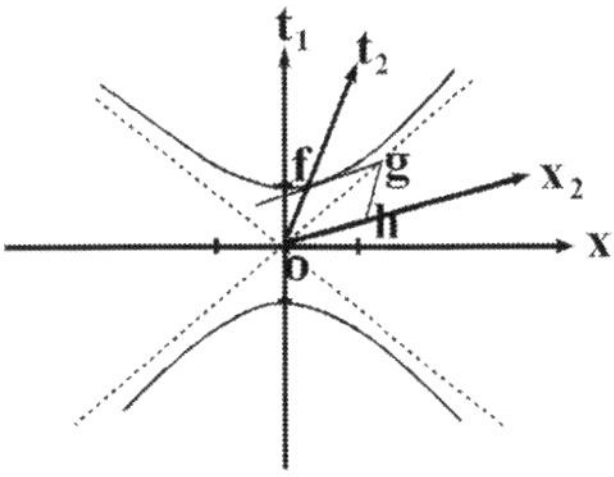

Figure 8.6. The Minkowski Diagram

Now it might be said that the fly, flying the length of the cigar lying along x_1, is flying a longer distance as far as Y is concerned since he determines X's space units have expanded. But the distance that the fly traverses in sixty wing-beats – however great or small the distance is *measured* to be – this distance holds a fundamental "causal flow" or invariant that relativity and its measurement procedures cannot alter. If we mark this distance by two markers, A and B, the fly will buzz from A to B in sixty wing-beats, no matter what the reference system from which he happens to be viewed. It is the "sixty wing-beat distance invariant." We start from this. The fly flies this distance every day, from the cereal bowl to the sticky spoon on X's table, in sixty wing beats. Relativity, simply because Y goes into motion, contains no inherent justification for altering this.

Assume that the rocket is moving at 80% the speed of light. Given Y's view of X as having contracted time units, the same sixty wing beats require 1.66 seconds as assigned to X by Y. So, now we partition this sixty wing beats (an invariant causal flow) across the 1.66 seconds. In X's normal system, at sixty wing beats/second, there are six wing beats in each $1/10^{th}$ second, and X can normally perceive or discriminate one wing-beat per $1/10^{th}$ second. Thus at six beats per each $1/10^{th}$ second, he sees a blur. In the new partition assigned by Y, with sixty beats partitioned over the 1.66 seconds, X sees only 3.6 wing beats in each $1/10^{th}$ second. It is less a blur. The fly appears to be buzzing more slowly. X's time (his perception of the rate of events) is slower, despite the fact that his velocity of processes has not changed. This is clearly absurd, yet this is exactly what is required of the world of X if we ignore reciprocity, and if these transformations are ontological enough to support Y's eventual return as more youthful than X.

On the other hand, there is the effect on Y, whose time units are expanded and space units contracted. In Y's moving system, a fly is buzzing across the table in the rocket cabin, again using sixty wing beats from A to B. It requires only .6 of the expanded Y-second for the distance to be covered. The invariant sixty wing beats are partitioned across this amount, therefore becoming ten beats per each $1/10^{th}$ second, and thus the fly is now more of a blur, despite the unchanged velocity of processes. It can be argued, just as Eddington notes, that due to the rocket's velocity, Y's processes are retarded. But in fact everything in Y's reference system is retarded, to include the fly and its buzzing from A to B. In effect, we have simply subtracted a constant across all motion values of the system, and the problematic modification of perception just noted still holds. In essence, psychology contradicts physics.

In this analysis, I have stayed consistently within the implications displayed in the Minkowski diagram, that is to say, within the case where Y is consistently the one in motion, X stationary. If we want to set X in motion, we need another diagram, and the situation simply reverses.

The Role of Reciprocity

What is wrong here? There is the strange picture of Y's view of X's altered perception of events in X's own system. But let us ignore this. One aspect of the problem is more elementary. As noted, when we represent the situation of X and Y in the Minkowski diagram, we have fixed on one observer, X, and set all other systems in motion relative to him. The Minkowski schema represents the adjustments in time and space units necessary to preserve light-velocity invariance for all other systems. But it cannot represent reciprocity. We could equally have fixed on Y and set all other systems in motion with respect to *him*. This, again, requires another diagram, and so on for each observer upon whom we fix.

Given this, we must ask the fundamental question: is the effect on either X or Y a real effect? Y, we know, could equally declare his system to be at rest, and X in motion relative to him. Clearly, the effects cannot be real from this perspective. The different "times" and "distances" represent only the observer's method of keeping his measurements consistent with light-velocity invariance. STR, from this perspective, fails to justify, either for X or for Y, a different perception of the fly based on the observer's motion. If we respect the inherent reciprocity of reference systems in STR, there is no contradiction with the relativity of perception. STR is at worst neutral with respect to a causal flow in time (the fly) invariant to both X and Y. Only if we insist that STR implies a real effect is there a contradiction.

It must be clearly understood again here that I am not denying the empirical facts, e.g., increase of life spans in mesons, or the retarded clock carried by the jet, or increases in mass. The empirical evidence is not in dispute. These are real effects. What is in dispute is the use of STR to explain the empirical evidence; it is used inappropriately in attempting to do so. The structure of reciprocity intrinsic to STR is being ignored.

Half-Relativity

"Half-relativity" is what Bergson termed the asymmetric use of STR.[15] The Lorentz equations are applied to the meson; the life span increase falls out via t'. End of explanation. As noted already, A. P. French, in a textbook that attempts to maintain clarity, in a section entitled "Relativity is Truly Relative," flatly states that the time dilation (just as the length contraction of the Michelson-Morley apparatus) as observed for a meson is not a property of matter but something inherent in the measurement process.[16] He goes to the rare extent of actually showing *two* Minkowski diagrams, one for each observer (as though there were a small observer on the meson), to show the symmetry of the changes in *each* system. Just as Bergson argued earlier, French notes that were an observer to compute t' as the meson falls to the earth, the tiny observer on the meson is equally allowed to say that he is stationary and the earth moving towards the meson. This is to say we have here, in French's terms, a "measurement effect." Thus, when French treats the twin paradox, he invokes the *asymmetry* introduced when the twin on the rocket turns around to return, therefore introducing a new inertial frame.[17] STR is used to compute the different (shorter) "time" of the traveling twin for each leg of the trip, thus ascribing the magnitude of the difference to v. But he assumes, in conjunction with this, that it is the asymmetry introduced by the turn-around that is required to support the real (aging) effect, i.e., as *a real property of matter*. Clearly, if one twin is now gray and has a long beard, we have a change that is a real property of matter. Thus he argues that STR, factoring in this asymmetry associated with the turn-around and its acceleration, and due to the fact that a time difference value can be derived due to v, can indeed handle the twin paradox. Yet he has earlier painstakingly built the case, to the point of doubled Minkowski diagrams, that the structure of STR demands symmetry (reciprocity), and given this symmetry, it does *not* explain any changes as real properties of matter.[18] In essence, the entire explanatory burden for aging as a real effect now falls on the asymmetry introduced by the change in inertial frame. *But where is this theory!?? That is,* where *is the theoretical framework supporting how and to what magnitude introducing an asymmetry affects the physiological processes underlying aging? Or why the asymmetry can be introduced into STR?* More precisely, where is the theory that explains how introducing an asymmetry now allows the use of the Lorentz equations independent of, or outside of, the symmetric, reciprocal structure provided in STR?

In the comparison between X and Y above, with our invariant fly, we only asked Y to be in uniform relative motion at velocity v, just as in the meson case, just as in the Michelson-Morley case. This comparison could care less about Y's return or differential accelerations. We don't need a rocket. While X sits by the kitchen table watching the fly, Y could travel by on his tricycle, and the same relativistic laws hold.[19] Nevertheless, there are those that would simply classify this case as the twin-paradox, invoke the existence of accelerations, and move the problem and the effects involved into the General Theory. All of the effect can then be assigned to acceleration(s). This reaction is extremely problematic. If we seize upon any accelerating component of a motion (which one can always find, even for the startup of the tricycle) to allow us to get to the safety of the GTR, then what if anything is the province of STR? The physics would be in danger of becoming a shell game, shuffling an explanatory pea between STR and GTR. If we are doing this to avoid reciprocity, then the argument that STR, with its inherent reciprocity, fails to explain any of these effects is effectively conceded, and this lynchpin in its being a theory of time – its ability to explain these effects – is removed.[20] Note again, it is not the aging effect, it is *all* asymmetric effects – jet carried clocks or long living mesons – that would have to be so moved into GTR for consistency. One dismisses the above comparison of X and Y into the GTR then only with difficult consequences.

Thus others (as well as French) have argued, as Eddington appeared to believe, that the twin-effect is perfectly consonant with STR. But to stay fully within the context of the Special Theory without bringing in gravitational field changes, Salmon envisaged a rocket ship (A) departing earth and passing another (B) coming in the opposite direction at the same velocity. At the point of meeting, the two exchanged signals to coordinate their clocks.[21] B continued on to earth where clocks were compared, and of course, in a triumph for the theory, an earthbound observer's clock showed a greater passage of time than B's. This appears to be ironclad, yet there is a problem. Reciprocity has not been avoided. The observer in A takes with him his own reference system. Since no reference system is privileged, he has equal right to declare himself at rest and everything else in motion relative to him, including the earth, the earthbound observer, and the earthbound observer's clock. When B passes A and signals are exchanged, will they then reflect a decrease in the rate of A's time? Hardly, given A is at rest. Only the author of the argument happens to believe A is in motion, but he forgot to ask A.[22]

The twin-paradox is disturbing precisely because it epitomizes, very concretely, the inconsistency relative to standard use of STR. It highlights a very real effect, e.g., a youthful man versus a hoary old one, that cannot simply be assigned to a measurement process. Interestingly, Einstein himself, in a (little known) 1918 article, attempted to preserve reciprocity and the asymmetrical effects together by arguing that indeed the rocket ship could be considered stationary, its motors only neutralizing the pull of the earth as the earth recedes.[23] But he then argued that it would require such tremendous field changes to move the earth and bring it back that the earth twin would undergo rapid aging. The reciprocity and the paradox denying the reciprocity appear resolved (just as French argued). But now, ignoring the ad hoc, physically unrealizable fields, it is not clear of what use relativity is here at all. Its mathematics, with its intrinsic reciprocity, now does not accurately describe the phenomenon – we can clearly distinguish the two systems via gravitational effects – and it would seem logically prior to have a theory relating gravitational changes to a model of the physiological processes driving aging – this in itself being sufficient to account for the phenomenon without appealing to changes of "time" itself. The one-way application of the Lorentz transformations would then appear in retrospect to be but a convenient empirical description of these events, but a deeper theory would provide a model of the processes involved (as Lorentz himself attempted).

The Half-relativity of 1905

Einstein, for all practical purposes, began assigning real effects due simply to v, ignoring reciprocity, in the very paper that introduced the theory, in 1905. In the paper, he quickly invokes the reciprocity implied in the first postulate, having us envisage a rigid sphere of radius R, at rest in the moving system.[24] At rest relative to the *moving* system, he notes, it is a sphere. Viewed from the "stationary" observer the equation of the sphere's surface gives it the form of an ellipsoid, with the X dimension shortened by the ratio $1:(1 - v^2/c^2)^{1/2}$. He notes (the reciprocity) then immediately: "It is clear that the same results hold good of bodies at rest in the 'stationary' system, viewed from a system in uniform motion." Two paragraphs from this point he notes the "peculiar consequence" that were there two synchronous, separated clocks A and B in the stationary system, and if A is moved to B with velocity v in time t, it will lag behind B by $\frac{1}{2} tv^2/c^2$. The structure of reciprocity is already being voided here – we are dealing only with an effect in the stationary system, not relating the two systems. The observer in the stationary system can simply move the clock from A to B to fulfill Einstein's condition, and the effect is simply ascribed to v (Figure 8.7). This conclusion is quickly reinforced. Within another paragraph, Einstein, extending this to "curvilinear motion," states flatly that this result implies that a clock at the equator must go more slowly, by a small amount, than one situated at the poles, i.e., again two clocks in the same system, in this case the earth.[25] Physicists accept this equatorial clock retardation naturally as a real effect. The effect had to be factored in to Hafele and Keating's jet-carried clock experiment. Yet reciprocity demands that the clock on the equator be stationary, the observer at the pole spinning around. Now it is not a real effect. This is likely not very tasteful. Yet this conclusion regarding v as already producing real effects in 1905 is doubly reinforced when it is considered that the equator-clock is an exact analogue to Einstein's future thought experiment (introducing GTR) of the rotating disk. Now the observer leaves the center of the disk, moving along a radius to the rim and back, while carrying a clock. Upon his return the clock is retarded.

The thought experiment used this result as a very real effect. Yet why? The observer takes with him, at every point he occupies, his own proper time. He should return with the clock unchanged.

The "Comfort" of the GTR

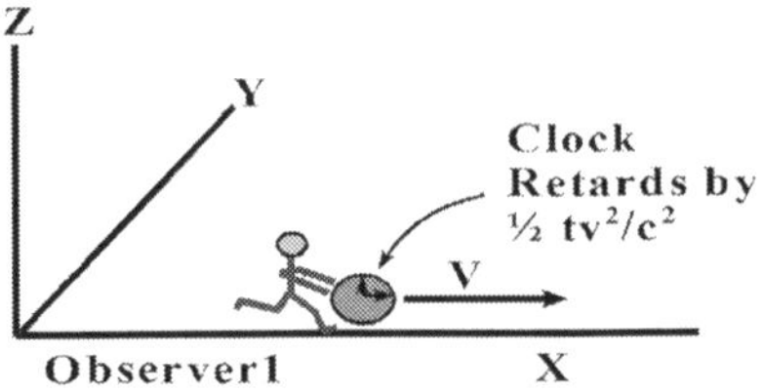

Figure 8.7. The observer causes the clock to retard simply by moving it at velocity v within his own reference system.

It is neither my intention nor scope here to delve deeply into the General Theory. It is invoked so often as an escape valve for these problems that something more should be said that gives the escapees at least a little pause. The issue, further, goes to the heart of the origin of these real effects. The comfort of assigning these real, ontological changes to the GTR arises from the tenet that acceleration breaks the symmetry or reciprocity of systems. Because, as A. P. French did for example, we can invoke acceleration as being involved when the rocket turns around and heads back to earth, we can fix the retarded aging on the acceleration as the cause. Acceleration is considered *real or absolute*; it is not subject to the reciprocity of reference systems as is uniform motion (velocity). That is, it is not the province of STR.

Acceleration was seized upon as an equivalent to gravity by Einstein in his thought experiment on riding in an elevator. In a uniform motion up, I notice nothing, but if the elevator accelerates, I feel a force. I can know I am moving. The motion is not relative, nor subject to STR. This principle sits at the heart of GTR. Acceleration is a real force. One can question Einstein's conclusion. Bergson argued simply that acceleration cannot be distinguished from velocity in the sense relativity claims – velocity is a rate of change in position over time, acceleration simply the rate of change of the rate of change of position. That is, one is the first derivative of change in position with respect to time, the other is simply the second derivative. Why is the second derivative so privileged over the first? Physicist, Ling Jun Wang, refines this argument, deriving the generalized Lorentz equation for t' in the context of acceleration.[26] If we cannot integrate over infinitesimal velocities, he argues, as did Bergson also argue, we have undercut all of physics. Wang's equation completely undercuts any appeal to the GTR due to acceleration in the twin paradox; in fact it implies a question to the foundation of GTR. As I noted in Chapter V, the source of GTR's foundation via this principle is in the attempt to use *force* to explain the *real* motion of the matter-field. Simultaneously it represents the inability of the classic metaphysic to come to grips with the primacy of motion.

Why is the problem of "real effects" significant? There are three reasons. Firstly, if STR is being used inappropriately as an explanatory device where the one-way use of the mathematics just happens to work, then physics should be searching for the true explanation. It could be extremely instructive, if only for the apparent return of the ether, which formerly housed some of these effects (again, in Lorentz's mind for example), in more sophisticated form as the quantum vacuum.[27] Secondly, there is now the contradiction with the psychology of perception just discussed and which I hope would merit at least some review. Thirdly, if we cling to the idea that STR can explain real, asymmetric effects, then we are equally clinging to the reality of the relativization of simultaneity, i.e., to the *real* breakup of simultaneity into successive moments in time, and vice versa. It is this implication that I wish to further question.

The Relativity of Simultaneity

In Figure 8.8 we picture three points, A', B', and C' in Y's moving system placed along the direction of this motion. Each will be a distance L from each other. We will assume Y is at point B', and the system is moving with velocity v. From the viewpoint of the stationary X, these three events are not simultaneous. The clock at A' registers a time slightly behind that of B', while the clock at C' is somewhat ahead. The greater the value of v, the greater this lag and lead time respectively. Both times are given by Lv/c^2 seconds (where c is the speed of light, L is the distance of each clock from the center of the system). As v approaches the speed of light c, the maximum time difference becomes L/c seconds (distance / velocity = time, or L/c = time).

If we drop a perpendicular from A' to K', this line will symbolize all the past events at A'. Since we see that the clock is slow at A', and Y then supposedly looking at past events, this line displays the maximum reach into this past. Likewise the line upwards from C' to H' shows the maximum of the future. Now we can draw yet another line of simultaneity, this one running to (hypothetical) points D' (between C' and H') and E' (between A' and K'). Its divergence from the original line A'B'C' is a function of the speed v. Further, were the difference in v between the X and Y systems infinitesimally small, there would be a line barely divergent from A'B'C' representing the fact that at *even the*

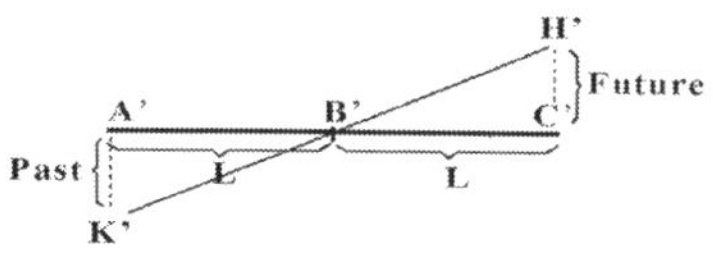

Figure 8.8 Planes of simultaneity

most infinitesimal velocities, we see the breakup of simultaneity begin, radiating from the most minute point or distance from B', increasing in degree towards A' and C'. There is any number of such lines.

What is the reality here? Imagine that Y is moving at an infinitesimally small velocity relative to X. For practical purposes, X's line ABC and Y's line A'B'C' are virtually coincident. But yet, even at the most minute velocity, simultaneity has begun to break up at the most infinitesimal point or distance from B, increasing in degree as we approach A' or C'. Now Y moves at a much higher velocity. X now notes the difference in Y's clocks. He is forced to assign events at A' deeper and deeper into Y's past as v increases, and to assign events at C' farther into the future. He does this by the very fact that he needs to keep the velocity of light invariant as per the Lorentz transformations. But Y can equally say he is at rest. He continues to note the simultaneity of events at A', B', and C'. He now notes the same breakup of simultaneity for X. Again the question becomes, is the conversion of simultaneity to succession real? Is it more than a notational convention required for the consistency of measurements between the two systems? Can this possibly be true of the flow of time?

Rakić's Critique

Perhaps we are asking here, though relativity has hitherto withstood the test of time, whether time has truly been used to test STR. Natasa Rakić was able recently to show that there is a very reasonable feature of time that STR must incorporate, yet is not definable in terms of the causality relation inherent in Minkowski space-time.[28]

Consider Figure 8.9. The left side depicts the light-cone concept of relativity. The lines of the X are rays of light. These rays show the path of the fastest possible causal action (light speed) that could emanate, so to speak, from an event. Any thing/event falling within the top "V" of the X can be influenced or causally affected by the event. There is a flip side, or bottom of the X. Anything inside the bottom of the X can influence or causally affect the event. On either side of the X is the "causal elsewhere" of the event. Any happening here cannot affect the event. It would have no causal relation to the event.

On the right, three events (e's) are pictured. The dotted "X" is the light cone of e_2. Thus event e_3 lies within the top V of e_2's "X" – e_2 can causally influence e_3. Event e_1 also lies within the top V of e_2's "X". So e_2 can causally influence e_1. Event e_3 however is in the causal elsewhere of e_1. It cannot causally influence e_1. To make this a little more concrete: At event e_2 an

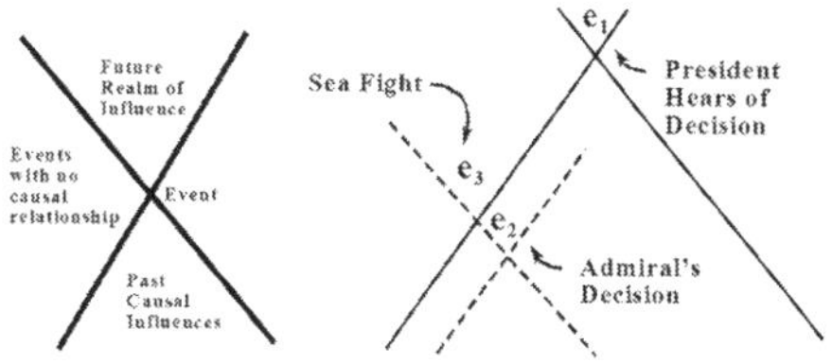

Figure 8.9. Left, the "light-cone" of an event. Right, the admiral's decision (e_2), the sea fight (e_3), and the president's learning of the decision (e_1).

admiral decides to engage the enemy ships, an event (the battle) which subsequently occurs at e_3. While the sea battle at e_3 is in the causal future of e_2, for e_1 it is in the area of events with no causal relationship. At e_1 the president learns of the admiral's decision (e_2). Pretending we are dealing with events moving at light speed here, we can easily imagine a movie scene where Roosevelt is consulting with his advisors, wondering what is going on in the Pacific theatre, and being told of Admiral Spruance's decision to engage at Midway. The entire scene takes place in this pregnant context of meaning, for we know simultaneously that the brave admiral and his fleet have already sailed into the fray. Thus, Rakić noted that it is clearly possible that e_3 (the sea battle) has already become, i.e., is *realized* with respect to e_1 (when the president hears about it). Though at e_1 the message about the battle has not yet been received, this does not imply that the sea battle has not occurred, i.e., is not realized. By this is meant that there is a very real, *ontological* relationship between e_3 (the battle) and e_1 (learning about the decision). The relativistic model of time is entirely insufficient to represent the actual fabric of time.

Rakić is content to term past, present and future not temporal but ontological relations, letting STR hold the ground as a theory of time. She notes however that her arguments could be construed as saying that STR is not a theory of time, allowing ontology and time to go together as is usual to common sense, but argues that in this case common sense must concede.

Holding Rakić's strained concession in place is undoubtedly STR's apparent success as an explanatory theory, all of which unfortunately revolves on asymmetric effects. But, the inadequacy of the Minkowski schema to represent time and its causal flows goes more deeply. When the flight of the fly, with his sixty-wing-beat distance invariant, was treated as an invariant causal flow, the beginning of the problem for the relativization of simultaneity was already introduced.

The Simultaneity of Flows

The intuition of a universal flow is partially preserved in relativity in the conservation of a "causal order." On analysis, we will find multiple causal orders or flows within this flow as Bergson noted or even, as Gibson insisted in the opening quote, where hero rushes to save the endangered heroine. The simultaneity of flows is integrally bound to causal order and to a global transformation wherein the motions of "objects" are transferences of state. Consider two football players running down each sideline of the field at precisely equal velocity. A physicist (O_1) at the fifty yard line notes the time against two synchronized clocks on each sideline as the players run by and ascertains that they have passed the same point simultaneously (Figure 8.10, e_1 and e_2). Of course a second physicist (O_2), thinking the first in motion and noting this observation says the first is in error, the events were not simultaneous. Yet the two football players continue on, converging on a football equidistant from both that they both kick simultaneously (e_3), kicking the ball twice as far as just one would have achieved. From the perspective of an instantaneous measurement, i.e., abstract time, their simultaneity is relativized. From the perspective of the two causal flows, the simultaneity of the flows is absolutely real. The second physicist cannot deny the effect of the simultaneous kick. One cannot simply relativize multiple causal flows.

It can be argued that e_1 and e_2 are not truly simultaneous just as O_2 states, that simultaneity is achieved only at the point-instant of the kick. But we could replace the football players equally well with a huge cue stick sweeping down the field towards a billiard ball. Positioned at each yard line are O_1's measurement clocks. If the cue's outside edges truly fail to pass the measurement clocks/ points simultaneously, it will hit the ball at a slant sending it off at an angle (Figure 8.11). In sliding the x_1, x_2 and t_2 axes upwards towards e_3, it can be seen that there will come a point as our very wide cue nears the ball at e_3, that e_3 will fall in the causal elsewhere of the light cones of each of the edges (e_1, e_2). This implies that the two outer edges could not possibly be squared in time for a flush contact of the entire cue surface with the ball if they are as non-simultaneous as claimed by O_2. The global causal flow led by the cue's frontal surface is fragmenting under STR's treatment. Yet the cue strikes the ball precisely perpendicularly. Only one strand in this flow, one local flow, the causal order in STR invariant to both observers, is ultimately preserved. This is the chain of causal relations, <, the relation determining time-like and space-like events, defined upon a sequence of infinitely minute point-instants extending through the time line t_1 to e_3. Were we considering the fly, no matter how infinite the "points" we place on this

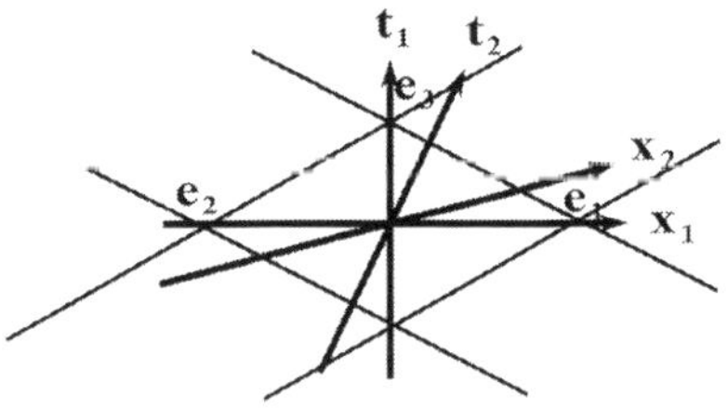

Figure 8.10. Two football players (e_1, e_2) converge on the ball (e_3).

line, or the in fact multiple lines comprising the fly, this will remain sixty wing beats – an indivisible movement or flow. A global flow, whether fly or cue stick or hero and heroine, cannot be an invariant to all observers in STR.[29]

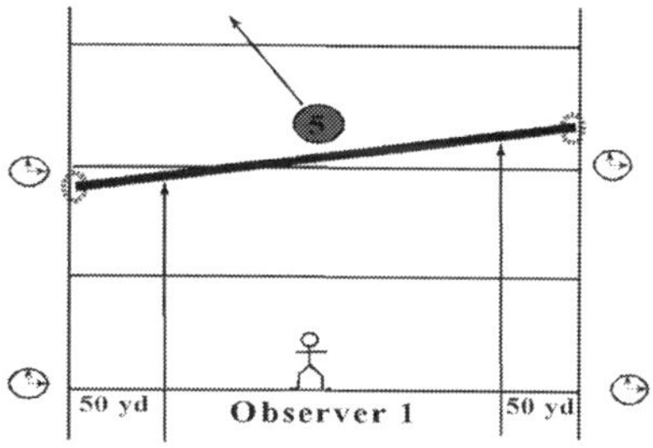

Figure 8.11. The flow of the very wide cue stick up the football field. If its edges are not simultaneously passing the yard lines, as Observer 2 believes, it must strike the ball at a slant.

We must ask what is the causal validity or efficacy of this one local point-instant flow? The breakup of simultaneity, as we have earlier seen, drives downwards in space-time to the most infinitesimal of point-instants. At this mathematical point, as earlier noted, there is neither time nor events. As such, without the possibility of even an event, it is impossible to say that there is anything causal whatsoever with respect to this point, or with respect to a "causal" chain of such points. The abstract space and abstract time that support the classical concept of causality offers again an infinite regress. If this chain is infinitely divisible – an infinite set of "point-events" – then between each point we must introduce a "causal relation," which is in effect to say a motion ad infinitum. Causality too will require indivisible extents. The fly, as a coherent biological system doing his sixty wing-beat trip, is precisely a global, indivisible flow. Were he taking his sixty wing beat trip to e_3, the tips of his wings will stop precisely simultaneously, O_2's measurements to the contrary notwithstanding. When it was insisted earlier that this sixty wing beat flow be treated as invariant to both X and Y, this weakness inherent in STR's treatment emerged.

In the above, I have not attempted a formal definition of a causal flow. I am leaving this at the intuitive level where, for example, a fly, as a complex system in motion, is comprised of multiple processes acting in concert, be this multiple muscle systems, neurons firing, or chemical flows. Such a system could be as large, and larger, as a weather system such as a hurricane, or an evolving galaxy, or a collection of individuals all working together to play a symphony. The two football players with which we began were two seemingly isolable local flows. They could, however, have been two sailboats moving in unison before a vast pressure front. Or this could have been a vast magnetic flux sweeping the earth. The point is that we must ask if any such local flows, any more so than "objects" and their "motions," are truly isolable from the global time-flow of the universal field. Are they more than transferences of state within the global motion? This global transformation is the classical "flow of time" invariant to all observers.

STR and Consciousness

Two theorists, Hagan and Hirafuji, analyzed the concept of the "emergence" of consciousness in the context of relativity.[30] Emergence envisions consciousness arising (or being generated from) from the physical processes in the brain, analogous, it can be said, to the glow arising from the filament of a light bulb. That is, it is the same old "experience is generated by the brain" notion. Their analysis deals a critical blow to the emergence concept, but a deeper reading indicates that doubt is cast on STR's ability to support of any theory of consciousness.

Starting with what they term the *extrinsic* definitional problem, they argue that any emergent state of consciousness (experience) must be frame invariant to satisfy the requirement that the conscious state be invariant to another observer in motion. What this means, let me immediately interject, is simply that the simultaneity of all the musicians playing a symphony that one is watching better be maintained! Hagan and Hirafuji aver that keeping the emergent property frame invariant might be achieved, but choose not to explore the difficulty, moving on to yet another extrinsic problem (what they term a "boundary" problem). In fact, it cannot be achieved. Our experience, we have seen, is marked by the characteristic of simultaneity of flows – the multiple melody lines within a single flow of a symphony, multiple musicians playing on the symphony stage, multiple women cooking in the kitchen, etc.

From the standard view of relativity, from which Hagan and Hirafuji write, the simultaneity of any of the above systems (read *experiences* as well) should indeed breakup, simultaneity becoming succession, and succession becoming simultaneity. Recall the three points, A', B', and C' of Figure 8.8, and the break up of simultaneity at the most infinitesimal difference of velocity. We asked if this can possibly be true of the flow of time? In the more obvious causal context of causal flows, e.g., our two football players, we saw that this cannot be true. One cannot simply relativize multiple causal flows.

Yet this is precisely what relativity would do. Each of the experiences mentioned earlier, with their simultaneous flows, would begin to breakup relative to the motion, for example, of observer Y. This is why the "emergent" consciousness or emergent "property," as Hagan and Hirafuji mention, would have such difficulty remaining frame invariant. More correctly, this is why the invariance is impossible. The experience would inevitably be distorted relative to the frame. But as I asked earlier, can we seriously believe this "breakup" of succession and simultaneity is possible, i.e., *that it has any ontological status!*? Do we believe the symphony would become jumbled, the musicians playing out of time, the conversations at the table scrambled, the cooking women putting ingredients in the cake one after the other rather than together, etc.?

One could question the relevance of the frame invariance requirement. So what, if from Y's point of view, my consciousness is distorted? It is my consciousness and it is perfectly OK, the symphony is fine, the ladies' conversation is fine. But this is the problem: if the theory (STR) is taken to indicate that this distortion would indeed be so from Y's perspective, i.e., it has ontological status, despite the intuitive oddity of the claim, we must ask what good is the theory? Hagen and Hirafuji are not only demonstrating the difficulties with a theory of "emergence" in the context of current physical theory, but also the difficulties for relativity of supporting any model of consciousness.

Let us move to the *intrinsic* definitional problem. Hagan and Hirafuji show that an intrinsic definition, while not requiring simultaneity, will always be incompatible with locality constraints. The difficulty here stems from the transmission speeds of the brain or, simply the very need or constraint for finite transmission. Under these constraints, the brain could not support a global state underlying an emergent property. The global state cannot inform the local dynamics of the boundary necessary to establish the physical extent of the emergent unit. But in essence here, I note, we have come back to the need for simultaneity, for this is an essential feature of any emergent property of consciousness or perception of which we can conceive.

Stein, as we saw, attempted to explain ongoing misconceptions of relativity, as he saw them, in terms of our continued naïve belief in the perception of simultaneous events – an illusion based on the high velocity of light.[31] Thus, he argued in essence, the naïve or intuitive simultaneity that perception provides is founded upon the "fleeting motions" of "masses of elements" in the brain, all subject to the limitation of communication via the velocity of light and implying, therefore, that at a small enough scale of time, perceptive simultaneity would break down. Stein is assuming a model of the processes in the brain underlying perception. But it is precisely this "fleeting motion" of masses of "elements" that Hagan and Hirafuji demonstrate is subject to locality constraints and, in being so subject, cannot support the simultaneity inherent in conscious states or perception, at least not from an "emergence" standpoint. If we only require a classical dynamics within the brain, under the locality constraint, to support a specifying reconstructive wave, as per Bergson's model, we escape the emergence difficulty, but this framework, with its non-differentiable time and simultaneity of flows, leaves relativity behind.

The Non-Relativistic Brain

While by the very nature of the abstract space or point continuum, motion in the continuum, as we have seen, is intrinsically relative, yet, the growth of a tree, taken, say, as a set of points, is a system of simultaneities – a simultaneous causal flow. No arbitrary part of this motion can be relativized – declared at rest or motionless relative to the remainder of the moving (growing) system. Nor can the motion of the whole of this flow be relativized – it cannot be declared at rest (halted) relative to some other system in motion. It is a *real*, not relative motion. One can model this motion or growth via the abstract space, but this continuum has properties that tell us it is both meaningless to the real motion of the tree and not capturing the actual reality. Nor, we would quickly discover, is the tree as an object actually isolable from the rest of the dynamic flow of the universe.

The abstract space with its continuum is just as meaningless to the operation of the brain as it is to the growth of the tree. The brain, like the growing tree, is not actually operating in the abstract space/continuum. As a physical device, the brain knows *nothing* about the continuum, it does not exist to the brain. The continuum is a conceptual derivative, a resultant from cognitive processes of the brain that a conscious being applies conceptually to external reality and its events, unfortunately often as though this abstract continuum were indeed real. It is even applied by conscious beings to the brain and its operations, as in the computer model, which, in strong AI, is a complete creature of belief in the *reality* of the abstract space and abstract time, but this application is utterly wrong. *The brain does not work "in the abstract continuum,"* i.e., the continuum as defined by the classic metaphysic. It is working in an entirely different, yes, metaphysical structure. Therefore the brain cannot be modeled, or better and more importantly – *built or reconstructed* – as a creature of the continuum of the classic metaphysic.

Some theorists have attempted to drive the problem of consciousness down to implementing the brain's actual biology as the critical issue, but in this stance, this entire question of a metaphysic is being invoked. The brain as a biological system is the tree. If you are implementing a working brain as a biological system – assuming you know what the biological system actually does and how it functions (particularly in perception) – you are implementing a device that functions in a different (temporal) metaphysic. Your mathematical models may approximately describe what it is doing, but they do not actually capture what it is doing. And if one thinks one can design a robot, even though the robot is a real physical, dynamic machine, but yet – for its operations supporting vision, for perception, for thought and cognition – is only executing syntactic operations in an abstract space and an abstract time (that is simply another dimension of this space), one is missing this point.

The Singular Time of Consciousness *and* Physics

There have been other examinations of STR, of both its explanatory status in physics and as a theory of time. Bergson was perhaps the earliest. His argument in *Revue Philosophique* with physicist Andre Metz circa 1924 centered on the use of STR in explaining asymmetric effects.[32] Metz could neither accept that STR is an inappropriate explanatory vehicle, nor could he conceive of the possibility that the increased life spans of mesons could be explained without resorting to STR. Deleuze would reprise Bergson's general argument on time with respect to relativity.[33] Dingle would make interesting critiques, particularly on the invariance of light.[34] Brillouin would give a non-relativistic explanation of the retardation of atomic clocks (and of the red shift).[35] Earman would note that there has yet to be a relational, let alone a relativistic explanation of Newton's humble bucket.[36] Nordenson would argue that Einstein's rejection of the classical flow of time, whether beyond "proximity" or anywhere even beyond the mathematical point, must surely undermine any meaning to his new procedure for clock synchronization.[37] Rakić, in proving certain logical inadequacies of the Minkowski metric, is reduced to declaring Special Relativity to be not an ontological theory, but concedes it a status as a "temporal" theory. Whatever meaning this concession might have, a theory with no ontological status is of little use; it is certainly not relevant to a science of perception or a theory of consciousness. Nehru, describing Dewey Larson's alternative theory of time, clearly sees the questionable status of these real effects in relativity per se and the need to derive them far more organically. Note the "so-called:"

> …All [Einstein's] so-called Relativistic effects come out, in the Reciprocal System, of this existence of this additional time component.[38]

For anyone accepting and using relativity in their thought on space and time, or trying to integrate it with quantum theory, you must ask, *just what theory are you using*? If it is the theory that is supposedly supporting relativistic effects as ontological or real, it is an utter mess of contradictions, and begun so by Einstein himself. This is not to mention the problems then generated by the implied reality of relativized simultaneity. If it is the true *Invariantentheorie* with its self-consistent use of invariance under transformation, it is useless as a theory to explain such effects as ontological, yet we know these exist. I cannot comprehend how one can fail to ask this "W*hat theory*?" question. Further, it is a theory that cannot support the obvious reality of simultaneous causal flows and it is a theory that is simply an extension of an outmoded metaphysic that will never support consciousness, perception or the quality of the perceived world. These are absolute requirements for a theory of time.

STR, with its confused interpretation and its reflection of the classic, spatial metaphysic with its degenerate view of "time" is an impediment to both physics and psychology. Physics has struggled to both reconcile STR/GTR with quantum theory (aggravated by the awareness of quantum theory's non-locality) and simultaneously to understand and perhaps incorporate the role of consciousness in quantum theory. The theory of time is precisely the ground where psychology, the theory of consciousness and physics meet. In truth, with Bergson's vision of time – with its non-differentiable flow, with its irreversibility derived from the fact that each "instant" reflects the entire preceding series, with its primary memory or true continuity wherein there are no mutually external "instants," where the motions of "objects" are transferences of state within a global time-evolution of the material field implying therefore an inherent non-locality – one sees that Einstein's two times, "a psychological time different from that of the physicist," are in reality one.

Chapter VIII: End Notes and References

1. Gibson, the highly respected theorist of perception, made this statement in a talk at the University of Minnesota in 1975. He had read a paper by the author the previous day which at the time accepted Capek's (1966) view that relativity adequately preserves the "becoming" of the universe, and which attempted to fold in psychological time as part of the relativistic structure of time. Gibson, however, correctly appeared to have none of this, i.e., none of relativity whatsoever. He is in effect alluding to the concept of the *simultaneity of flows* of time, a subject discussed at length by Bergson in *Duration and Simultaneity* (1922/1965) in his analysis of relativity.

   Capek, M. (1966). Time in relativity theory: Arguments for a philosophy of becoming. In J. T. Fraser (Ed.), *The Voices of Time.* New York: Brasiller.
2. This chapter is a modification of the article below:

   Robbins, S. E. (2010). Special Relativity and Perception: The Singular Time of Psychology and Physics. *Journal of Consciousness Exploration and Research,* 1, 500-531.
3. Gunter, P. A. Y. (1969). *Bergson and the Evolution of Physics.* University of Tennessee Press., pp. 123-135.
4. Gunter, P. A. Y. (1969). *Bergson and the Evolution of Physics.* University of Tennessee Press, p. 133.
5. Stein, H. (1991). On relativity theory and openness of the future. *Philosophy of Science, 58*, 147-167.
6. Davies, P. & Gribbons, J. (1992). *The Matter Myth.* New York: Simon & Schuster, p. 100-101.
7. Horton, G. (2000). *Einstein, History and Other Passions.* Cambridge, Massachusetts: Harvard University Press.
8. French, A. P. (1968). *Special Relativity.* New York: Norton, p. 114.
9. Smythies, J. (2003). Space, time and consciousness. *Journal of Consciousness Studies, 10*, 47-56.
10. Weyl, Hermann (1922/1952) *Space, Time, Matter.* Dover, p. 217 (emphasis added).
11. Eddington, A. (1966). *Space, Time and Gravitation.* Cambridge: MIT Press, p. 57.
12. Dunne, J. W. (1927). *An Experiment with Time.* London: Faber and Faber.
13. Smart, J. J. C. (1967) Time. *The Encyclopedia of Philosophy.* New York: Collier-MacMillan.
14. Smythies, J. (2003). Replies from John Smythies. http://tech.groups.yahoo.com/group/jcs-online/message/2582.
15. Bergson, H. (1896). *Matter and Memory,* p. 255.
16. Bergson, H. (1922/1965). *Duration and Simultaneity With Respect To Einstein's Theory.* Indianapolis: Bobbs-Merrill.
17. French, A. P. (1968), p. 114.
18. French, A. P. (1968), pp. 155-156.
19. I have been posed one objection or "solution" to this problem, yes, by a reviewer, stated as follows: "The twin leaving and returning on the rocket ages less because his worldline between departure and return is shorter. And the length of the worldlines is observer invariant." This is a strange misconception, mis-statement and convolution.

The "observer invariance" is only defined within the structure of symmetric (reciprocal) transformations created by both observers. There is no "invariance" with but one observer. But then it is this very symmetry that makes it impossible to use relativity to explain changes as real properties of matter.

20. Brillouin (1970) would argue that a reference system must be very massive to reduce all action-reaction effects. The tricycle, let alone an abstract "coordinate system," would not qualify in his opinion. The same point however can be made with a more massive system going by the table. But I do not believe that Einstein was concerned at all with this distinction, the geometry being the overriding consideration. Brillouin, L. (1970). *Relativity Reexamined.* New York: Academic Press.
21. Salmon, W. (1976). Clocks and simultaneity in special relativity, or, which twin has the timex? In P. K. Machamer & R. G. Turnbull (Eds.), *Motion and Time, Space and Matter.* Ohio State University Press.
22. Davies (1977) resolves the twin paradox by flatly assigning the aging differential to the turn around at the target star and the homeward acceleration of the rocket (pp. 43-44). Yet, like French, he applies the Lorentz equations, claiming that he has also preserved the symmetry, a fact his table of durations (p. 44) obviously belies, for only the rocket clock shows a consistent, time-expanded 4.8 light years for each leg – the rocket is clearly the only object moving to Davies. Davies (1995) drops the clear emphasis on acceleration as the root cause of the aging. He does declare "there is no paradox" because the symmetry is broken due to accelerations in the necessary stop and return of the rocket, but never mentions this again. Ignoring the consequent inapplicability of STR, he again proceeds to apply the Lorentz transformations (with what justification?). In essence, he notes that at 80% of the speed of light, earthbound twin Ann would see the clock of the rocket-twin (Betty) as running .6 of earth-Ann's. Symmetrically, rocket-Betty, viewing herself as stationary, sees earth-Ann's clock as running .6 of Betty's. This symmetry holds for each leg – the outward and the homeward bound. In Davies's scenario, it is rocket-Betty who returns having aged less, not earth-Ann, and he claims that he has resolved Dingle's (1972) critique that in this case, "each clock runs slower relative to the other," in other words, a critique which says precisely that there can be no ontological status here. Given the symmetry he took great pains to describe, Davies conveniently never tells us why earth-Ann does not also have the distinction of aging less.

    Davies, P. (1977). *Space and Time in the Modern Universe*. London: Cambridge University Press.

    Davies, P. (1995). *About Time: Einstein's Unfinished Revolution*. New York: Simon & Schuster.
23. A translation of this paper is discussed in Dingle (1972, pp. 191-200).
24. Einstein, A. (1905/1923). On the electrodynamics of moving bodies. In H. A. Lorentz, A. Einstein, H. Minkowski, H. Weyl. *The Principle of Relativity.* New York: Dodd Mead, section 4, p. 48.
25. Einstein, A. (1905/1923). On the electrodynamics of moving bodies. In H. A. Lorentz, A. Einstein, H. Minkowski, H. Weyl. *The Principle of Relativity.* New York: Dodd Mead, p. 50.

26. Wang, L. (2003). Space and time of non-inertial systems. *Proceedings of SSGRR 2003*, L'Aquila, Italy.
27. There are probably any number of ways, for example, to account for the life-span increases of mesons without resort to the mystical "changes of time" required by STR. Thomson's model of the electron, as just one possible example of an approach, saw the electron as a special case of an electric current. In motion, a current naturally generates a counter-EMF – a resistance to its own motion, a resistance increasing with velocity, unto a singularity. So too would a single electron. Now if the meson is a group of electrons and positrons, where the positrons radiate away the group's energy as a function of a certain synchrony, this being "decay," then putting the group in motion will retard this radiation, the decay rate ever decreasing with speed, and increasing its lifespan. (Cf. for example, Aspden, 1969, 1972; Kessler, 1962).
    Aspden, H. (1969). *Physics Without Einstein.* London: Sabberton.
    Aspden, H. (1972). *Modern Aether Theory.* London: Sabberton.
    Kessler, J. (1962). *The Energy of Space.* Published by the author.
28. Rakić, N. (1997). Past, present, future, and special relativity. *British Journal for the Philosophy of Science, 48*, 257-280.
29. A comment on concepts expressed by Myrvold is appropriate here. Myrvold considers the relation eRe' (where R = "realized with respect to") in the context of *extended* objects. This requires taking a spacelike slice – in effect an instantaneous stage along some foliation of the object's history. Failure to do this results, he notes, in paradoxes like the "pole and barn," where, with the barn at rest and the pole in high velocity motion through the barn, there is a period where the pole just fits inside the barn, and conversely, with the pole at rest, and the barn in motion, there is no such "pole-inside" state. This conflict is resolved, he argues, "by remembering that the states of the extended system of which one account speaks are states along spacelike slices of the system different from those of which the other account speaks" (p. 478).

    This is a not a justifiable modification of STR. The reference system of Figure 8.8 would be treated as a set of points, α. Another set, β, would be definite or realized with respect to α if in α's causal past. Though seemingly applying to the cue stick example, we could not extend the system indefinitely, or it would extend across the entire universe, providing a plane of simultaneity. But, given Myrvold, what prevents this move? My earlier analysis relative to Figure 8.8 shows that the simultaneity of α begins to break up at the most minute interval relative to an observer in motion. But, there is a simpler reason why Myrvold is not a resolution. If the length contraction of the pole is being taken as a real effect in this paradox, the (very testable) implication is that we could actually trap the pole inside the barn, different spacelike slices or not. Such a real result (captured pole) is as much a contradiction as the twin paradox. If it is not considered a real (possible) effect, this is due to giving the reciprocity of reference systems its appropriate status, which is to say there is no ontological status to the relativistic contraction, and no "paradox" in the first place. Myrvold dismisses the paradox, considering it an example of misunderstanding, yet it is no more a misunderstanding than the twin-paradox where the "time-change" should have equally as little ontological status.

    Myrvold, W. (2003). Relativistic quantum becoming. *British Journal of the Philosophy of Science, 54*, 475-500.

30. Hagan, S., & Hirafuji, M. (2001). Constraints on an emergent formulation of conscious mental states. *Journal of Consciousness Studies*, *8*, 109-121.
31. Stein, H., op.cit, 1991.
32. Gunter, P. A. Y. (1969). *Bergson and the Evolution of Physics.* University of Tennessee Press, pp. 135-190.
33. Deleuze, G. (1966/1991). *Bergsonism.* (Translated by H. Tomlinson and B. Habberjam) New York: Zone Books.
34. Dingle, H. (1972). *Science at the Crossroads.* London: Martin Brian & O'Keeffe.
35. Brillouin, L. (1970). *Relativity Reexamined.* New York: Academic Press, pp. 77-85.
36. Earman, J. (1989). *World Enough and Space-time.* Cambridge: MIT Press.
    Newton noted how, when a bucket filled with water is spun rapidly, the water is flung (by the centrifugal force) in a standing wave pattern against the sides of the bucket. How, he asked, can this motion/force of the water be relativized with respect to the bucket? It is an absolute motion or effect.
37. Nordenson, H. (1969). *Relativity, Time and Reality.* London: George Allen and Unwin.
38. Nehru, K. (2009). Precession of planetary perihelia due to coordinate time. Reciprocal System Theory Library, March 16, 2009. http://library.rstheory.org/articles/KVK/Prec-PlanetPeri.html

CHAPTER IX

Epilogue

If at the very centre of my universe there had been a blob – a little and tightly packed and intensely personal boxful of neural and material processes – how crazy to suppose that such a puny thing could meaningfully encompass the Cosmos and its origin and the whole mystery of Being!

- Douglas E. Harding, *On Having No Head*

For many people the mind is the last refuge of mystery against the encroaching spread of science, and they don't like the idea of science engulfing the last bit of *terra incognita*.

- Raymond Kurzweil quoting Daniel Dennett quoting Herbert Simon, *The Age of Spiritual Machines*

I have no sooner commenced to philosophize than I ask myself why I exist... When finally a Principle of creation has been put at the base of things, the same question springs up: How – why does this principle exist rather than nothing?

- Bergson, *Creative Evolution*

Other Realities

There is a remarkable study by Dr. Rick Strassman (*DMT: The Spirit Molecule*) on the effects of DMT. DMT is the much more powerful cousin of LSD. The experiences tend to fall in a few major categories. One category relates to the body and one's life circumstances. A second type invariably involves a certain limited set of realities or realms – basically two – beyond our normal experience, while a third category is extremely close to the experiences reported in the literature on Near Death Experiences or NDEs. In this latter, there is a sub-category that involves far higher mystical states than that normally reported in the NDE literature.

One of the realms in the second category involves experiences, often in lab-like settings, with beings perceived as alien forms. No other drug Strassman is aware of opens this latter type of experience, or better, tunes us to this realm. The experiences are remarkable. Once injected with DMT, within fifteen seconds the person feels a rush and suddenly finds himself perceiving a completely different environment, with no major alteration in the quality of awareness, and usually there appear one or more "beings" in this environment who interact with the person and are felt, with certainty, to be entirely "real" entities, independent of, but not exactly separate from, the DMT tripper's mind. As one of Strassman's subjects describes his experience:

> ...I felt like I was in an alien laboratory, in a hospital bed like this. . .. A sort of landing bay, or recovery area. There were beings. . . .They had a space ready for me. They weren't as surprised as I was. It was incredibly un-psychedelic. I was able to pay attention to detail. There was one main creature, and he seemed to be behind it all, overseeing everything. The others were orderlies, or dis-orderlies. ...They activated a sexual circuit, and I was flushed with an amazing orgasmic energy. A goofy chart popped up like an X-ray in a cartoon, and a yellow illumination indicated that the corresponding system, or series of systems, were fine. They were checking my instruments, testing things. When I was coming out, I couldn't help but think, "aliens."[1]

It is obviously difficult to ignore the possible relation here to the experiences of alien abductions, or to the phenomena experienced in NDE's (Near Death Experiences), or to the many varieties of experiences in astral worlds, or to any number of related phenomena. This is why Strassman's thoughts are interesting here. As he explains, it was the consistently similar experiences from DMT with what could only be identified as "aliens" that he found so remarkable. Rather than assign them to that easy, but illusory explanatory bin, "generated by the brain," he argued for something quite different.

DMT is produced naturally in the body by the pineal gland. Strassman speculated that certain meditative practices might stimulate the release of DMT, leading to spiritual experiences in other realms. He also felt that the phenomenon of alien abduction is so similar to certain DMT trips that they're likely the same thing. This however does not diminish the "reality" of the experience. Strassman uses the TV analogy, where we tune to other channels. He argues:

> ...DMT provides regular, repeated, and reliable access to "other" channels. The other planes of existence are always there. . . . But we cannot perceive them because we are not designed to do so; our hard-wiring keeps us tuned in to Channel Normal. It takes only a second or two – the few heartbeats the spirit molecule requires to make its way to the brain – to change the channel, to open our mind to these other planes of existence."[2]

The brain, we have seen, is a reconstructive wave passing through the holographic, universal field. It is a wave supported by the concrete dynamics created by the biological, chemical, physical architecture of the brain. It is naturally, yes, "tuned" to Channel Normal – the ecological world at our scale of time, the scale of "buzzing" flies and stirring spoons. But the complex biochemical structure supporting this attunement can clearly be modulated for different attunements, just as we speculated in the context of introducing a catalyst to change our scale of time. When, as a result of the catalyst(s), we see the fly flapping his wings like a heron, the heron-fly is not being "generated by the brain," but we do have a new attunement to a different frequency of the events in the matter-field.

DMT is apparently a different form of modulation. If injected with DMT, and I consistently find myself in an alien dimension, and in some particular experience resting on a lab table, yet, the experimenter still sees me, motionless, sitting in the chair. My perception in this new dimension is not virtual action in our ecological world. I am not acting physically in this other dimension. Thus, there are other aspects of being implied here of which we have little understanding, some of which certainly revolve around the pineal gland itself.

The pineal, roughly the size of a pea and located in the center of mass of the brain, is now being seen to have very remarkable properties. It has connections to the visual system and it contains proteins normally involved in phototransduction, i.e., it has retinal like properties. Further, as David Wilcock notes in his review of this subject, it contains an array of hexagonal microcrystals similar to crystals in the inner ear, and it is possible that these crystals act piezoelectrically, transducing electromagnetic waves to light or photons.[3] In all respects then, this could be the "inner keyboard" that Bergson thought must exist, which is symmetric to the initial visual processing area (V1) that responds to an *external* object. This inner keyboard responds initially to *virtual objects or events in time*, allowing these to then be integrated within the dynamic processing of the brain and thus experienced as images.[4] I did not dwell on this aspect of Bergson in our discussion of redintegration as it does not change the emphasis on the invariance structure of events as being the key to redintegration. However, such a mechanism is implied in my comments on memory retrievals in the "Is Everything Stored" section, the discussion of the role of images in voluntary action, Libet's delay of the experienced intent in preparing an action, and Bergson's "circuit" integrating past memories into current perceptions, e.g., the "bear" in the woods that turns out to be only a tree stump. It is certainly a component of the complex dynamical state supporting explicit memories which require that "articulated simultaneity" relating past events and the present.

Despite this difference in terms of virtual action, if our model of perception in Channel Normal effectively bans us, as we have seen, from pronouncing that perceptual experience is simply "generated by brain," we must be equally unwilling to declare perception in other realities as simply "generated." This simplistic conceptual escape route is no longer available. Further, when we truly weight the fact in our models of the brain that perception is the result of concrete biochemical dynamics and fields, not simply symbol manipulation and abstract computations, we have an entrée in theory to studying these modulations. In the robotic model, there is no room whatsoever, no place, in fact no need, to allow for modulators such as DMT, or, for that matter, time-scale altering catalysts. We must entertain, then, that there are many other forms of modulation possible, many other attunements, attunements to what may be many levels of frequency and existence in the vast, universal field. We have only begun to explore the possibilities stretching before us when we unshackle ourselves from our limited, robotic, machine conception of the mind and brain.

Yes, as a software engineer for forty years, I am far from the illusion that I have created detailed specifications for the "device" I have sketched. We have the elements of a framework: The holographic field, the non-differentiable flow of time, the reconstructive wave, the invariance laws that describe the ecological world and allow for gauge (time-scale) invariance, the true relation of subject and object in terms of time, the bare essentials for the specification of scales of time, the principle of virtual action, the model of redintegration, the nature of analogy. But I am also far from the illusion that if cognitive science and neuroscience fail to use this framework, refuse to explore how the neural dynamics actually implements a reconstructive wave, that we will ever advance in our understanding of the brain, mind and consciousness. If these sciences choose to remain in the great Abstraction with its outmoded metaphysic and outmoded physics for that matter, they will wander aimlessly for years, announcing ever new, but blindly uncritical progress for their neural nets, while robots, seemingly becoming ever more capable, will continue to be used as iconic images, both to marginalize ESP, NDEs, perception in other realities and in general anything that doesn't fit the received view, and to cement this limiting conception into the human psyche.

Final Reflections

There is another email in my mailbox. The email announces the formation of a new journal, *The International Journal of Machine Consciousness*. I must sigh. What better way to cultivate this spurious concept within the culture than to create an academic journal on the subject? The journal description announces that, "Since the topic of machine consciousness is still highly controversial, each issue will endorse a blend of papers covering provocative theories as well as testable models." Note the, "*Still* highly controversial." By the gradual attrition of time, and the studied neglect of competing theories, the subject will undoubtedly become non-controversial.

Here is another email, this one from the JCS (*Journal of Consciousness Studies*) online forum I like to follow. The writer is noting an article in a journal, *Trends in Neuroscience*, describing a complex "associative memory" model. This model, I know, is nothing more than a version of the connectionist neural networks we have already discussed. But the *Trends* author is apparently invoking the ever threatening vision of a huge breakthrough in computing power. In 2015, there should be a supercomputer with a processing capacity of 10 petaflops or more (a petaflop is a thousand trillion operations per second). By 2023, if the standard pace of development holds, the forum writer speculates, there should be a computer with the capacity to simulate the real-time "associative memory" in the human brain. Unfortunately, it does not matter how many flops there are. Simulating a ton of logical operations – syntax rules – at light speed means nothing. As far as posing as a brain, the machine will still be a megaflop. The brain is an entirely different type of device; it is a concrete reconstructive wave within the non-differentiable flow of the holographic matter-field. But the sufficiency of computations never seems to loose its hypnotic grip on its worshipping subjects. Remember: the fields and dynamics of the brain are as concrete as an AC motor. It helps break the spell.

It is time to gaze out my window, stare at the field. Maybe the coyote is there… In the early 1990s, there was finally some more general realization in philosophy that the computer model might not be able to explain consciousness. Many new journals were created, even this very *Journal of Consciousness Studies*. I took this avenue to publish many chunks of the theory laid out here and its critique of the computer model. Though I felt these articles might have nudged the nose of the academic community into the profound and entirely ignored problem of time in relation to mind, there has been little if any attention. Probably to be expected. By and large, the theorists are still children of Plato and heirs to the same philosophical community that resolutely buried and still ignores Bergson. I noted earlier that in this *Journal of Consciousness Studies*, in an issue devoted entirely to machine consciousness, I was shocked to discover that the majority of theorists in the issue have so little grasp of the problem that they argued that even current robots might be conscious, or at least were within a hair of being so. If so, with just a little DMT, the robots will soon be visiting the aliens in their labs.

The robotic conception of mind demeans the spiritual nature of man. It demeans man in the work place. It debilitates education. It demeans man, period. When, according to Kurzweil's *The Singularity is Near*, his predicted "singularity" arrives in 2045, the robots that humanity will have created will supposedly be indistinguishable from man, in fact, they will well exceed man and our turtle-like speed of brain processing compared to the near light-velocity of the machines. By 2045. Forget creating mousetraps from a set of components and the deep problems of representational power this presents. This means that a robot will take granola bars, cherry tomatoes, Chinese Five Spice and piloncillo, and whip up a dessert in twenty minutes, easily becoming the champion of all *Chopped* champions.[5] Maybe even the *Iron Chef*. But, oh, maybe this would be yet another "unfair" extension of the Turing Test. Never mind, these robots will be unable yet to even see the coffee cup on my desk. But this is not understood by the Kurzweils of the world. Remember, to Kurzweil, the robotic machines will even have spiritual experiences.

Mystical experience is, nevertheless, experience. We have seen that AI and robotics are woefully unable to explain even normal experience. In fact, as we have seen, our normal experience of the coffee cup is just as mystical as the mystical experience of Zen. Normal experience is simply experience (the time-scaled, time-flowing field of matter) viewed through the "dynamical lens" of the great Abstraction – abstract space and time – with its population of separate "objects," to include oneself as one of these. This form of experience, strained through the Abstraction, is "Samsara," and this is why the Buddha would say, "Samsara *is* Nirvana." Turn off the dynamical lens creating Samsara; we find we are already in Nirvana.

There was a time, not so long ago, when physics was seen as naturally supporting the mystical. Fritjof Capra's *The Tao of Physics*, and *The Dancing Wu Li Masters* of Gary Zukav, as examples, were widely read and praised. What has happened since then? What has happened is the constant drumming of the implications for the nature of mind and brain derived from neuroscience, cognitive neuroscience, robotics and AI, combined with a complete failure to grasp the implications of Gibson and ecological psychology, and combined with the ongoing mythology of, or misconstrual of, the explanatory achievements of evolutionary biology. This completely spurious description of brains and evolution, as it were, seems to have superseded the lower level description of physics. But it is less than that. Messrs. Dawkins, Dennett and Kurzweil (my symbols for spokesmen for this view) have attacked from the weakest points of science – evolution and consciousness. In the first, evolution, a model of the actual mechanisms of evolution is still only a promissory note. In the latter, there is no model of consciousness. Cognitive Science has no clue how we see the coffee cup on the table. The attack is with a Styrofoam sledgehammer.

We cannot use science to turn humans into mere machines. We cannot use science to turn the universe into a mere machine. We are far from justified in either move. There is far more to this universe. There is much more to understand. We are children yet before it. I have argued that the universe clearly has an elementary awareness at the null scale of time. The properties of its indivisible, non-differentiable motion reside beneath our memory and our will. But what is the time scale of this universe? Our brain controls *our* scale of time; it gives us "buzzing" flies as opposed to "heron-like" flies. But the matter-field is an infinity of scales and retains the entirety of its past – trillions of lifetimes of beings. What do we really know of the nature of such a universe or ourselves as integral aspects thereof? It is far too early to limit our learning based on an unjustified conclusion from the weakest points of science.

Ray Kurzweil sees us being increasingly unable to distinguish machine intelligence from human intelligence. I would say, we will learn increasingly what human intelligence actually is. Mr. Kurzweil also chides us, warning skeptics against limiting the possibilities of his robots and the future for them he envisions. He quotes Arthur C. Clarke: "When a scientist sees that something is possible, he is certainly right. When he states that something is impossible, he is certainly wrong." To feel guilty of skepticism on whether robots can reach the limited conceptions of mind held by current Cognitive Science is absurd. It is the robotic conception of mind that is limiting our progress in exploring and understanding the dimensions and capacities of the human mind. I say that *this* is possible: experience is not stored in the brain, the brain accesses this experience as a reconstructive wave, perception is direct, memory is direct, the brain has access to an infinite realm of mind, mind infinitely overflows the brain. If we assume the impossibility of these statements, then who is that limits? We shall see who is right – and who it is that sees.

Chapter IX: End Notes and References

1. Strassman, Rick, (2001). *DMT: The Spirit Molecule*. Vermont: Part Street Press, p. 197.
2. Strassman, 2001, pp. 315-316.
3. Wilcock, D. (2011). *The Source Field Investigations*. New York: Dutton.
4. Bergson, 1896/1912, p. 129.
5. *Chopped* is a show on the Food Channel of cable TV where extremely accomplished chefs (and to me, probably genius chefs) are routinely given such challenges in competition with other chefs and the efforts are judged by experts. It is doubtful that Kurzweil grants much weight to this as an aspect of human intelligence, at least as little as building mousetraps.

29436412R00167

Made in the USA
Lexington, KY
25 January 2014